# Industrial Wastewater Treatment

# Industrial Wastewater Treatment

## NG Wun Jern
*National University of Singapore*

Imperial College Press

*Published by*

Imperial College Press
57 Shelton Street
Covent Garden
London WC2H 9HE

*Distributed by*

World Scientific Publishing Co. Pte. Ltd.
5 Toh Tuck Link, Singapore 596224
*USA office:* 27 Warren Street, Suite 401-402, Hackensack, NJ 07601
*UK office:* 57 Shelton Street, Covent Garden, London WC2H 9HE

**British Library Cataloguing-in-Publication Data**
A catalogue record for this book is available from the British Library.

**INDUSTRIAL WASTEWATER TREATMENT**
Copyright © 2006 by Imperial College Press

*All rights reserved. This book, or parts thereof, may not be reproduced in any form or by any means, electronic or mechanical, including photocopying, recording or any information storage and retrieval system now known or to be invented, without written permission from the Publisher.*

For photocopying of material in this volume, please pay a copying fee through the Copyright Clearance Center, Inc., 222 Rosewood Drive, Danvers, MA 01923, USA. In this case permission to photocopy is not required from the publisher.

ISBN  1-86094-580-5
ISBN  1-86094-664-X  (pbk)

Editor: Tjan Kwang Wei

Typeset by Stallion Press
Email; enquiries@stallionpress.com

Printed in Singapore by Mainland Press

## PREFACE

Students and engineers new to industrial wastewater treatment have often posed questions regarding the subject which may be answered from experience gained during multiple field trips. Organizing such site visits can however, be a difficult task because of time-management issues as well as the difficulties in gaining access to the various factories. This book was written to address some of these questions and to substitute a few of the site visits. It is a discussion of the material that goes into industrial wastewater treatment plants, the reasons for their selection and, where appropriate, how things may go wrong. Many photographs have been included so that the reader can get a better feel of the subject matter discussed.

Typically, students and engineers who wish to pursue a career in wastewater engineering begin from the study of domestic sewage and the design of sewage treatment plants. Their studies would then most likely extend to municipal sewage, which is a combination of domestic, commercial, raw and pretreated industrial wastewaters. Following which, some of these students may be briefly introduced to industrial wastewater treatment but their exposure to the subject would unlikely be of the same level as that of domestic sewage. Indeed, much of the expertise in the subject is gained through work experience. Many engineers, at least early in their careers, attempt to use the sewage treatment plant template or a modification of it for an industrial wastewater treatment plant.

How different is industrial wastewater treatment from sewage treatment? Is there a need to highlight the differences? Would these differences be large enough to result in differences in conceptualization, design and operation of industrial wastewater treatment plants? What are the potential pitfalls engineers should be aware of? There are obviously lessons to be learnt in sewage treatment which are relevant to industrial wastewater treatment. There is then the issue regarding the amount that can be transferred and the considerations that need to be taken into account to ensure an appropriate design is generated and the plant successfully managed.

Industrial wastewaters can be very different from sewage in terms of their discharge patterns and compositions. Notwithstanding this, many industrial wastewater treatment plants, for example and like sewage treatment plants, use biological processes as key unit processes in the treatment train. Given the variations in wastewater characteristics, ensuring these biological processes and upstream/downstream unit processes are appropriately designed presents a great challenge. The problems intensify when information on the wastewaters and their treatment is lacking. Textbooks frequently emphasize on the theories and equations used in designing unit processes. However, industrial wastewaters are so varied that it is difficult for the aspiring engineer to imagine why a certain process is selected over another, or why a particular variant is even selected at all. Additionally, there is a scarcity of books on "Asian" wastewaters and treatment facilities, which seems to be incongruous to the growing demand for Asia-focused books because of Asia's rapid economic development.

This book is intended to introduce the practice of industrial wastewater treatment to senior undergraduate and postgraduate environmental engineering students. Practitioners of the field may also find it useful as a quick overview of the subject. The book focuses on systems that incorporate a biological treatment process within the treatment train, with the material of the book largely drawn from the author's practice and research experiences. It does not delve into the details of theory or the "mathematics" of design, but instead discusses the issues concerning industrial wastewater treatment in an accessible manner. Some prior knowledge of the theory behind the unit processes discussed and the manner in which they are supposed to work is assumed. A description of a typical sewage treatment plant is provided to afford readers a point of familiarity and basis for comparison so that the differences can be more apparent. The book approaches the development of suitable treatment strategies by first identifying and addressing important wastewater characteristics. In the latter part of the book, a number of specific wastewaters are identified to serve as case studies so that individual treatment strategies and plant concepts can be move clearly illustrated.

Ng Wun Jern
April 2005

## CONTENTS

Preface                                                                 v

**Chapter 1 — Introduction**                                            1

*Discussion on the impact of industrial wastewater discharges on the environment with a focus on Asia.*

**Chapter 2 — Nature of Industrial Wastewaters**                       12

*Discussion on a number of the key industrial wastewater characteristics which may impact on plant design and successful plant operation. Tables showing the characteristics of wastewaters arising from a variety of industries are included. A portion of this information is on wastewaters not usually found outside of tropical or sub-tropical regions. It is intended this chapter becomes a reference for professionals seeking information on wastewaters.*

**Chapter 3 — The Sewage Treatment Plant Example**                     28

*Brief description of the possible treatment trains in a sewage treatment plant — based on the continuous-flow bioreactor and cyclic bioreactor. This is intended to provide a framework for comparison so that readers can more readily appreciate the differences and similarities between sewage treatment plants (STPs) and industrial wastewater treatment plants (IWTPs).*

**Chapter 4 — The Industrial Wastewater Treatment Plant — Preliminary Unit Processes**     42

*Discussion on the preliminary treatment required to prepare industrial wastewaters for secondary treatment. This chapter includes discussions on removal of suspended solids, O&G, inhibitory substances, pH adjustment, nutrients supplementation, and equalization.*

## Chapter 5 — The Industrial Wastewater Treatment Plant — Biological     61

*Discussion on the biological processes used for secondary treatment of industrial wastewaters to remove organics and nutrients (where necessary). Aside from discussion on aerobic processes such as the conventional activated sludge and the cyclic SBR, space is also devoted to anaerobic processes used as the first stage of a biological treatment train to reduce organic strength prior to aerobic treatment. The difficulties faced by biological processes in industrial wastewater treatment are highlighted.*

## Chapter 6 — The Industrial Wastewater Treatment Plant — Sludge Management     99

*The preliminary and secondary treatment stages generate sludges. These may be organic, inorganic, or a combination of the two. This chapter discusses sludge management approaches commonly adopted at IWTPs.*

*Chapters 4, 5 and 6 draw on experiences with actual wastewaters to illustrate points made in the discussions. These three chapters and Chapters 7–10 are provided with numerous photographs of plants, equipment, and site conditions so that the reader can develop a "feel" for the issues inherent in industrial wastewater treatment.*

## Chapter 7 — Chemicals and Pharmaceuticals Manufacturing Wastewater     106

*The pharmaceutical wastewater example provides a framework for discussion on the importance of segregation and blending, and the impact of inhibition.*

## Chapter 8 — Piggery Wastewater     112

*The piggery wastewater example provides a framework for discussion on the necessity to note the differences in wastewaters which may arise because of differences in industry practices (between Asia and Europe in this instance) and the approach taken to deal with high concentrations of SS in a highly biodegradable wastewater.*

## Chapter 9 — Slaughterhouse Wastewater     125

*The slaughterhouse wastewater example provides a framework for discussion on the importance of pretreatment to reduce a nitrogenous oxygen demand so that total oxygen demand may be reduced. Failing this the strong nitrification may require alkalinity supplementation with attendant implications in terms of treatment chemicals and construction materials needed.*

## Chapter 10 — Palm Oil Mill and Refinery Wastewater    134

*The palm oil mill wastewater example provides a framework for discussion on the use of anaerobic processes to treat wastewaters and not as is usually encountered in STPs to treat sludges.*

## References    145

## Index    147

# CHAPTER 1

# INTRODUCTION

## 1.1. The Backdrop

In many parts of the world, economic, social and political problems have arisen following rapid industrial development and urbanization, resulting in adverse effects on the quality of life. Urbanization in general initially places pressure on and overstrains public amenities. However, long-term and wider issues would eventually also be encountered as industrialization and urbanization exert pressure on the larger resource base that supports the community. This larger resource base includes forestry, freshwater and marine resources, as well as space suitable for further development. The difficulties associated with environmental degradation often originate from industrial development. They are amplified by rapid urbanization that is responsible for the growth of many major cities. In Asia, urbanization is exacerbated by large rural–urban migrations. These migrations emerge in response to perceived opportunities for a better livelihood in industrialized, economically booming urban areas. Rapid industrialization and its concentration in or near urban centers have placed very high pressures on the carrying capacity of the environment at specific locations. At these locations waterbodies such as rivers, lakes, and coastal waters have typically been severely affected.

Freshwater is a vital natural resource that will continue to be renewable as long as it is well managed. Preventing pollution from domestic, industrial, and agro-industrial activities is important to ensure the sustainability of the locale's development. Undoubtedly the water pollution control efforts which have been underway in many countries have already achieved some success. Nevertheless the problems that are confronted grow in complexity and intensity. It is estimated that 785 million people in Asian developing countries have no access to sustainable safe water (Sawhney, 2003). The pollution of freshwater bodies with the consequent deterioration in water quality can only worsen the situation. Such pollution has been brought about by the discharge of inadequately treated sewage and industrial wastewaters. This book focuses on the latter. Perhaps not unexpectedly, as the demand for more water is met, the volumes of wastewater can also be expected to increase. Coastal waters are also under pressure as they receive effluents discharged directly into them or indirectly from rivers. While most communities in Asia do not use coastal waters as a source of potable water (via desalination), there is already a movement towards this direction, as in the case of Singapore. Even though coastal waters are not yet a major source of potable water, they are, nevertheless, very important since they support fisheries and tourism industries. The ecosystems in many of Asia's coastal waters are fragile; damage to these ecosystems as a result of pollution can adversely affect fishery industries. The latter, in many instances, depend on mangrove forests as spawning grounds for marine life which are subsequently harvested.

Industrial wastewaters (including agro-industrial wastewaters) are effluents that result from human activities which are associated with raw-material processing and manufacturing. These wastewater streams arise from washing, cooking, cooling, heating, extraction, reaction by-products, separation, conveyance, and quality control resulting in product rejection. Water pollution occurs when potential pollutants in these streams reach certain amounts causing undesired alterations to a receiving waterbody. While industrial wastewaters from such processing or manufacturing sites may include some domestic sewage, the latter is not the major component. Domestic sewage may be present because of washrooms and hostels provided for workers at the processing or manufacturing facility. Examples of industrial wastewaters include those arising from chemical, pharmaceutical, electrochemical, electronics, petrochemical, and food processing industries. Examples of agro-industrial wastewaters include those arising from industrial-scale animal husbandry, slaughterhouses, fisheries, and seed oil processing. Agro-industrial wastewaters can be very strong in terms of pollutant concentrations and hence can contribute significantly to the overall pollution load imposed on the environment. It is perhaps ironic that the very resources which promoted industrial development and urbanization in the first place can subsequently come under threat from such development and urbanization because of over and inappropriate exploitation. Appropriate management of such development and resources is a matter of priority. The South Johore coast was such a case (ASEAN/US CRMP, 1991). This was then, economically, one of the fastest growing areas in Malaysia and potential damage to the environment of such development, if not properly managed, was recognized.

The impact of industrial wastewater discharges on the environment and human population can be tragic at times. Some 50 years ago, the Minamata disease which spread among residents in the Yatsushiro Sea and the Agano River basin areas in Japan was attributed to methyl mercury in industrial wastewater (Matsuo, 1999). However, tragedies as dramatic as the Minamata episode have not occurred frequently. Nevertheless, instances of pollution with potentially adverse impacts in the longer term have continued to occur. In the interim before the realization of these longer term impacts, a decline in the quality of life arising from the deterioration in water quality which various populations must access may become increasingly discernable. Examples of these, their recognition, and the efforts made to remedy the situations in the 1980s include the protection of Malaysian coastal waters from refinery wastewater (Yassin, 1987), the Tansui River in Taiwan where pesticides and heavy metals were discovered in the sludge (Liu & Kuo, 1988), the Nam Pong River in Thailand which was polluted by the pulp and paper industry (Jindarojana, 1988), and the Buriganga River in Bangladesh which had been polluted by, among other industries, tanneries (Ahmed & Mohammed, 1988). Similar reports in the 1990s include the Kelani River in Sri Lanka (Bhuvendralingam *et al.*, 1998), the Laguna de Bay in the Philippines (Barril *et al.*, 1999), and the Koayu River which had occurrences of *Cryptosporidum* oocysts and *Giardia* cysts after receiving inadequately treated piggery wastewater (Hashimoto & Hirata, 1999). Such reports are still frequent in the 2000s and caused concerns in Vietnam (Nguyen, 2003) and Korea (Kim *et al.*, 2003). The fact that water pollution due to discharges of inadequately treated industrial wastewater has occurred over decades in Asia obviously means solutions have not been found for all such occurrences. Towards the end of 2004, the Huai River in China was reported to have been so seriously polluted by paper-making, tanning and chemical fertilizer factories, farmers in Shenqiu County had fallen very ill after using the river water (The Strait Times, 2004). There has, however, been progress and an example is the successful ten year river pollution clean-up program in Singapore (Chiang, 1988).

Agro-industrial wastewaters, as a sub-class of industrial wastewaters, can have considerable impact on the environment because they can be very strong in terms of pollutant strength and often the scale of the industry generating the wastewater in a country is large. Citing ASEAN countries in Asia as examples, agro-industrial wastewaters had and in some instances still contribute very significantly to pollution loads. For example in 1981 the Malaysian palm oi and rubber industries contributed 63% ($1460\,\mathrm{td}^{-1}$) and 7% ($208\,\mathrm{td}^{-1}$) of the BOD (Biochemical Oxygen Demand) load generated per day respectively. This is compared with $715\,\mathrm{td}^{-1}$ of BOD from domestic sewage (Ong *et al.*, 1987). In the Philippines, pulp and paper mills generated $90\,\mathrm{td}^{-1}$ of BOD load (Villavicencio, 1987). Agro-industrial sites are therefore often the largest easily identifiable point sources of pollutant loads. While there are exceptions, individual industrial wastewater sources associated with manufacturing in Asia are, in contrast, more often small to medium sized compared to the former. The classifications of a small and medium-sized manufacturing facility have been defined in terms of the numbers of employees employed at such sites — $10 \sim 49$ persons and $50 \sim 199$ persons respectively.

Notwithstanding their small to medium sizes, the collective contribution from such enterprises to pollution is not necessarily negligible.

It should also be noted that while industrial wastewater sources may be small to medium-size, they are generally located in urban centers where building congestion is already a problem. To aggravate the situation, such factory operations may have no long-range project planning and are also unable to exploit advantages associated with economies of scale. A number of such operations may also try to maximize profits by reducing overheads and "unnecessary" expenditure associated with pollution control requirements — the result of an absence of an appropriate corporate culture and hence a weaker social conscience in terms of care for the environment. On a positive note, however, economic development over the last few decades has enabled necessary managerial, financial, and technological capabilities to address problems of pollution and environmental degradation over the broad spectrum of factory sizes. There is also a growing realization that the cost (in terms of the human and economic costs) of cleaning up after the act is frequently more than preventing the pollution in the first place.

## 1.2. What is Industrial Wastewater?

To begin the discussion on industrial wastewater, it may be useful to compare industrial wastewater with domestic sewage since designers of wastewater treatment facilities often begin their careers and almost certainly their education in environmental engineering by looking at sewage and sewage treatment plants. The latter can then provide a familiar framework which the reader can use to compare industrial wastewater and its treatment.

Domestic sewage is wastewater discharged from sanitary conveniences in residential, office, commercial, factories and various institutional properties. It is a complex mixture containing primarily water (approximately 99%) together with organic and inorganic constituents. These constituents or contaminants comprised suspended, colloidal and dissolved materials. Domestic sewage, since it contains human wastes, also contains large numbers of micro-organisms and some of these can be pathogenic. Waterborne bacterial diseases that can be present in sewage include cholera, typhoid and tuberculosis. Viral diseases can include infectious hepatitis. Inorganic constituents include chlorides and sulphates, various forms of nitrogen and phosphorous, as well as carbonates and bicarbonates. Proteins and carbohydrates constitute about 90% of the organic matter in domestic sewage. These arise from the excreta, urine, food wastes, and wastewater from bathing, washing, and laundering, and because of the latter, soaps, detergents, and other cleaning products can be found as well. Domestic sewage has a flow pattern which typically shows two peaks — in the morning before the start of working hours and in the evening after the population has returned from work. Typically these hydraulic peaks would also become more distinct as the sewage flows considered come from smaller populations and consequently smaller sewer networks. Variations in sewage characteristics across a given community tend to be relatively small although variation across communities can be more readily detected. Notwithstanding these variations, the composition of domestic sewage is such that it lends itself well to biological treatment in terms of the availability of and balance between carbonaceous components and nutrients. The biodegradability of sewage can be estimated by considering its Chemical Oxygen Demand (COD) and the corresponding $BOD_5$ (5 day BOD), and is indicated by its COD:$BOD_5$ and $BOD_5$:N:P ratios. This would typically be about 1.5:1 and 25:4:1 respectively. The nitrogen, N, would typically be in the form of organic nitrogen and ammonia-nitrogen (Amm-N). Nitrates ($NO_3$-N) would not be expected to be present as conditions in the sewers would be such that nitrate formation is unlikely while nitrate degradation because of anoxic reactions is likely. The phosphorous (P) would be a combination of organic and phosphate ($PO_4$) forms. The pH of sewage would be within the range of 6–9 and this is generally considered suitable for biological processes. Examples of values of $BOD_5$, TSS (Total Suspended Solids), and TKN (Total Kjeldhal Nitrogen) which have been used for purposes of plant design are 250, 300 and 40 mg $L^{-1}$ respectively. As indicated earlier in this paragraph, sewage characteristics can vary across communities and a raw sewage $BOD_5$ of 500 mg $L^{-1}$ has been encountered.

Industrial (including agro-industrial) wastewaters have very varied compositions depending on the type of industry and materials processed. Some of these wastewaters can be organically very strong, easily biodegradable, largely inorganic, or potentially inhibitory. This means TSS, $BOD_5$ and COD values may be in the tens of thousands mg $L^{-1}$.

Because of these very high organic concentrations, industrial wastewaters may also be severely nutrients deficient. Unlike sewage, pH values well beyond the range of 6–9 are also frequently encountered. Such wastewaters may also be associated with high concentrations of dissolved metal salts. The flow pattern of industrial wastewater streams can be very different from that of domestic sewage since the former would be influenced by the nature of the operations within a factory rather than the usual activities encountered in the domestic setting. A significant factor influencing the flow pattern would be the shift nature of work at factories. These shifts may be 8 h or 12 h shifts and there can be up to three shifts per day. These shifts may mean that there can be more than the two peaks in flow seen in sewage and there may be no flow for parts of the day. Factories may operate five to seven days per week. A consequence of this can be the possibility of zero flow on days when a factory is not operating. In contrast to the narrower band of variation in the characteristics of domestic sewage within a community, industrial wastewaters can have very different characteristics even for wastewaters from a single type of industry but from different locations. The cause of these differences has much to do with the operating procedures adopted at each site and the raw materials used therein. To further complicate matters, wastewater characteristics within a factory can also vary with time because it may practice campaign manufacturing, or it may practice slug discharges on top of its usual discharges. Apart from these events which occur on a regular basis, there would be spillages and dumping which may occur within the factory infrequently but can have very adverse impacts on the performance of the wastewater treatment plant. Consequently it would be prudent to assess an industrial wastewater, as well as its pretreatment and treatment requirements very carefully and not immediately assume that its wastewater characteristics and treatment requirements are similar to a previously encountered example. It would also be prudent to acquire some understanding of the nature of the factory's operations. A more detailed discussion of the characteristics of industrial wastewaters is made in Chapter 2.

On some occasions industrial wastewaters are discharged into a sewerage system serving commercial and residential premises. Such a combination of wastewater streams is known as municipal wastewater and the quality of such a mixture of wastewaters can vary depending on the proportion of industrial wastewaters in it and the type of industries contributing the industrial wastewater streams. Usually the domestic and commercial components in municipal wastewater can be expected to provide some buffering in terms of the characteristics of the combined flow. This is then expected to enable the combined wastewater to be treated easily compared to the treatment of the industrial wastewater on its own. However, even where the option of discharging into a sewerage system is available, some degree of pretreatment is frequently required at the factory before such discharge is permitted. Such pretreatment may include pH adjustment to 6–9 and $BOD_5$ reduction to 400 mg $L^{-1}$ as being currently practiced in Singapore (Pakiam et al., 1980). This is to protect the receiving sewers from corrosion and also protect the performance of the receiving treatment plant from an organic substrate overload.

## 1.3. Why is it Necessary to Treat Industrial Wastewater?

All major terrestrial biota, ecosystems, and humans depend on freshwater (i.e. water with less than 100 mg $L^{-1}$ salts) for their survival. The earth's water is primarily saline in nature (about 97%). Of the remaining (3%) water, 87% of it is locked in the polar caps and glaciers. This would mean only 0.4% of all water on earth is accessible freshwater. The latter is, however, a continually renewable resource although natural supplies are limited by the amounts that move through the natural water cycle. Unfortunately precipitation patterns, and hence distribution of freshwater resources, around the globe is far from even. Where precipitation does fall heavily, there are often difficulties with storage because of space constraints. Furthermore the available freshwater has to be shared between natural biota and human demands. The latter, aside from direct human consumption, includes water for agricultural, urban, and industrial needs. Freshwater shortages increase the risk of conflict, public health problems, reduction in food production, inhibition of industrial production expansion, and these problems threaten the environment.

Freshwater shortages are, however, not only due to uneven distribution of freshwater resources and demand for freshwater but also, increasingly, due to the declining water quality in freshwater sources already in use. This declining water quality is primarily due to pollution. It should not be forgotten that in the wider context of resources associated with water, the marine environment is also included in the picture. While the latter was, in the past, primarily associated with the fisheries resource, it can also include tourism and the feed for desalination in the current context. Untreated industrial wastewaters would add pollutants into waterbodies — freshwater and saline. These receiving waterbodies, freshwater and marine, can include ponds, lakes, rivers, coastal waters, and the sea. It would be useful to bear in mind that pollutants introduced into a river or some other freshwater waterbody do eventually end up in the sea, the ultimate receptacle for waterborne pollutants if these are permitted to find their way through the environment unimpeded. An example of riverine pollution are the rivers flowing through urban and industrial areas such as Hanoi and Ho Chi Minh City in Vietnam picking up pollutants such as heavy metals and organochlorine pesticides and herbicides. These pollutants reach the sea eventually and therein threaten the fisheries (Nguyen *et al.*, 1995). On Hainan Island (Southern China), for example, industries such as sugar refineries, paper mills, shipyards, and fertilizer plants accounted for about half the total wastewater generated and reaching the sea. This had resulted in incidences of the red tide in Houshui Bay and an area northwest of the island (Du, 1995). Obviously then, inadequately treated industrial wastewaters discharged into rivers would not only affect the freshwater in these areas but also the receiving coastal and sea waters. Eventually coastal resources such as the mangrove and reef ecosystems, and thereafter fisheries would be affected. The discharge of inadequately treated industrial wastewaters can therefore have far reaching consequences. In the last decade, the emergence of industrial pollution has been identified as a trend in the coastal areas of Southern China, Vietnam, Kampuchea, and Thailand.

The effects pollutants have on the water environment can be summarized in the following broad categories:

(a) Physical effects — These include impact on clarity of the water and interference to oxygen dissolution in it. Water clarity is affected by turbidity which may be caused by inorganic (Fixed Suspended Solids or FSS) and/or organic particulates suspended in the water (Volatile Suspended Solids or VSS). The latter may undergo biodegradation and thereby also have oxidation effects. Turbidity reduces light penetration and this reduces photosynthesis while the attendant loss in clarity, among other things, would adversely affect the food gathering capacity of aquatic animals because these may not be able to see their prey. Very fine particulates may also clog the gill surfaces of fishes and thereby affecting respiration and eventually killing them. Settleable particulates may accumulate on plant foliage and bed of the waterbody forming sludge layers which would eventually smother benthic organisms. As the sludge layers accumulate, they may eventually become sludge banks and if the material in these is organic then its decomposition would give rise to malodours. In contrast to the settleable material, particulates lighter than water eventually float to the surface and form a scum layer. The latter also interferes with the passage of light and oxygen dissolution. Because of the former, these scum layers affect photosynthesis. Discharge limits on wastewater or treated wastewater discharges typically have a value for TSS such as $30\,mg\,L^{-1}$ or $50\,mg\,L^{-1}$. Many industrial wastewaters contain oil and grease (O&G). While some of the latter may be organic in nature, there are many which are mineral oils. Notwithstanding their organic or mineral nature, both types cause interference at the air-water interface and inhibit the transfer of oxygen. Apart from their interference to the transfer of oxygen from atmosphere to water, the O&G (particularly the mineral oils) may also be inhibitory. Unlike domestic sewage, industrial discharges can have temperatures substantially above ambient temperatures. These raise the temperatures of the receiving water and reduce the solubility of oxygen. Apart from this, rapid changes in temperature may result in thermal shock and this may be lethal to the more sensitive species. Heat, however, does not always have a negative impact on organisms as it may positively affect growth rates although there are limits here too since the condition may favor certain species within the population more than others and over time biodiversity may be negatively affected;

(b) Oxidation and residual dissolved oxygen — As suggested in the preceding paragraph, waterbodies have the capacity to oxygenate themselves through dissolution of oxygen from the atmosphere and photosynthetic activity by aquatic plants. Of the latter, algae often plays an important role. However, there is a finite capacity to this re-oxygenation and if oxygen depletion, as a result of biological or chemical processes induced by the presence of organic or inorganic substances which exert an oxygen demand (i.e. as indicated by the BOD or COD), exceeded this capacity then the dissolved oxygen (DO) levels would decline. The latter may eventually decline to such an extent that septic conditions occur. A manifestation of such conditions would be the presence of malodours released by facultative and anaerobic organisms. An example of this is the reduction of substances with combined oxygen such as sulphates by facultative bacteria and resulting in the release of hydrogen sulphide. The depletion of free oxygen would affect the survival of aerobic organisms. DO levels do not, however, need to drop to zero before adverse impacts are felt. A decline to 3–4 mg $L^{-1}$, which still means the water contains substantial quantities of oxygen, may already adversely affect higher organisms like some species of fish. If inhibitory substances are also present, then the DO level at which adverse effects may be felt can be even higher than before. The case of elevated water temperatures due to warm discharges is somewhat different. The elevated temperatures can affect metabolic rates positively (possibly twofold for each 10°C rise in temperature) but elevated temperatures also reduce the solubility of oxygen in water. This would mean increasing demand for oxygen while its availability declines. Because of the impact of DO levels on aquatic life, much importance has been placed on determining the BOD value of a discharge. Typical $BOD_5$ limits set are values such as 20 and 50 mg $L^{-1}$;

(c) Inhibition or toxicity and persistence — These effects may be caused by organic or inorganic substances and can be acute or chronic. Examples of these include the pesticides and heavy metals mentioned in the preceding section. Many industrial wastewaters do contain such potentially inhibitory or toxic substances. The presence of such substances in an ecosystem may bias a population towards members of the community which are more tolerant to the substances while eliminating those which are less tolerant and resulting in a loss of biodiversity. For similar reasons, an awareness of the impact such substances have on biological systems is not only relevant in terms of protection of the environment but is of no less importance in terms of their impact on the biological systems used to treat industrial wastewaters. Even successful treatment of such a wastewater may not necessarily mean that the potability of water in a receiving waterbody would not be affected. For example small quantities of residual phenol in water can react with chlorine during the potable water treatment process giving rise to chlorophenols which can cause objectionable tastes and odors in the treated water. Apart from the organic pollutants which are potentially inhibitory or toxic, there are those which are resistant to biological degradation. Such persistent compounds can be bioaccumulated in organisms resulting in concentrations in tissues being significantly higher than concentrations in the environment and thereby making these organisms unsuitable as prey/food for organisms (including Man) higher up the food chain. While some organic compounds may be persistent, metals are practically non-degradable in the environment;

(d) Eutrophication — The discharge of nitrogenous and phosphorous compounds into receiving waterbodies may alter their fertility. Enhanced fertility can lead to excessive plant growth. The latter may include algal growth. The subsequent impact of such growth on a waterbody can include increased turbidity, oxygen depletion, and toxicity issues. Algal growth in unpolluted waterbodies is usually limited because the water is nutrient limiting. While nutrients would include marco-nutrients like nitrogen, phosphorous, and carbon, and micro-nutrients like cobalt, manganese, calcium, potassium, magnesium, copper, and iron which are required only in very small quantities, the focus in concerns over eutrophication would be on phosphorous and nitrogen as quantities of the other nutrients in the natural environment are often inherently adequate. In freshwaters the limiting nutrient is usually phosphorous while in estuarine and marine waters it would be nitrogen. Treatment of industrial wastewater (or domestic sewage for that matter) can then target the removal of either phosphorous or nitrogen, depending on the receiving waterbody, to ensure that the nutrient limiting condition is maintained. Given the litoral nature of many nations in Asia, removal of nitrogen would likely be necessary if the wastewater contained excessive quantities. When the nutrient limiting condition is no longer present in the waterbody, and when other conditions such as ambient temperature are appropriate, excessive algal growth or algal blooms (e.g. the red tide) may occur. Apart from aesthetic issues, such algal blooms may affect the productivity of the fisheries in the locale. It should be noted that not all industrial wastewaters contain excessive quantities of nutrients, macro and micro. This deficiency, if there is, results in process instability and/or the proliferation of inappropriate microbial species during biological treatment of the wastewaters. Bulking sludge is a manifestation of such an occurrence. To address this deficiency, nutrients supplementation is required. The quantities used should be carefully regulated so that an excessive nutrients condition is not inadvertently created and these excess nutrients subsequently discharged with the treated effluent. In terms of BOD:N:P, the optimal ratio for biotreatment is often taken as 100:5:1 while the minimum acceptable condition can be 150:5:1;

(e) Pathogenic effects — Pathogens are disease-causing organisms and an infection occurs when these organisms gain entry into a host (e.g. man or an animal) and multiply therein. These pathogens include bacteria, viruses, protozoa, and helminthes. While domestic and medical related wastewaters may typically be linked to such micro-organisms (and especially the bacteria and viruses), industrial wastewaters are not typically associated with this category of effects. The exception to this is wastewaters associated with the sectors in the agro-industry dealing with animals. The concern here would be the presence of such organisms in the wastewater which is discharged into a receiving waterbody and diseases, if any, are then transmitted through the water. While many of these organisms can be satisfactorily addressed with adequate disinfection of the treated effluent and raw potable water supplies during the water treatment process, there are those which cannot be dealt with so easily. Two examples of such organisms, *Cryptosporidum* and *Giardia*, were identified in Sec. 1.1. These belong to the protozoa family. The difficulty is that the infected host does not necessarily shed the organism but is likely also to shed its eggs or oocysts. The latter can unfortunately be resistant to the usual disinfection processes. An outbreak of cryptosporidiosis, a gastrointestinal disease, would result in the hosts suffering from diarrhea, abdominal pain, nausea, and vomiting.

With the above effects in view, industrial wastewater treatment would typically be required to address at least the following parameters:

(a) Suspended solids (SS);
(b) Temperature;
(c) Oil and grease (O&G);
(d) Organic content in terms of biochemical oxygen demand (BOD) or chemical oxygen demand (COD);
(e) pH;
(f) Specific metals and/or specific organic compounds;
(g) Nitrogen and/or phosphorus;
(h) Indicator micro-organisms (e.g. *E. Coli*) or specific micro-organisms.

# CHAPTER 2

# NATURE OF INDUSTRIAL WASTEWATERS

In the previous chapter a large variety of industrial wastewaters was mentioned. There are those with constituents which are primarily inorganic and thus would not be suitable for biological treatment. The focus of this book is on those with quantities of organics requiring removal and where biological treatment is a viable treatment option. It is very important for the designer and operator of a wastewater treatment plant to have as much knowledge of the wastewater's characteristics as possible. This is to ensure a suitable plant design is developed and the subsequently constructed plant is appropriately operated.

Industrial wastewater characteristics which would require consideration include the following:

 (i) biodegradability;
 (ii) strength;
 (ii) volumes;
 (iv) variations and;
 (v) special characteristics which may lead to operational difficulties.

An appreciation of the industrial processes used at a site is also often useful for understanding the reasons why particular attributes are either present or absent and why the variations occur. In time the designer may be able to anticipate some of the potential difficulties a plant may experience by observing operations at an industrial site.

## 2.1. Biodegradability

For an industrial wastewater to be successfully treated by biological means it should have quantities of organics requiring removal and these (and any other constituents present in the wastewater) should not inhibit the biological process. The quantity of organics in a wastewater is indicated by the wastewater's $BOD_5$ and COD (dichromate) values. Since the BOD is the oxygen demand exerted by micro-organisms to degrade organics while the COD is that required to chemically oxidize organics without considering the latter's biodegradability (i.e. approximately equivalent to the total organics present), the difference between the COD and BOD values would provide an indication of the quantity (in a relative sense but not in absolute terms) of non-biologically degradable organics. Similarly then the COD:BOD$_5$ ratio can provide an indication of how amenable a wastewater is to biological treatment. Since the dichromate COD value would always be larger than the BOD$_5$ value in an industrial wastewater, the COD:BOD$_5$ ratio should always be greater than 1.

It has, however, been noted that wastewaters with COD:BOD$_5$ ratios of 3 or lower can usually be successfully treated with biological processes. COD:BOD$_5$ ratios of 3 or lower are encountered in many of the agricultural and agro-industrial wastewaters. Table 2.1.1 provides information on six poultry slaughterhouse wastewaters. All six cases practice blood recovery although they may not have done so to the same extent. All have also practiced recovery of feathers and again in varying degrees. The COD:BOD$_5$ ratios of five of the examples which ranged from 1.3:1 to 2.5:1 suggested that such wastewaters are easily biodegradable and this has been noted to be so at the treatment plants. Case-5 had much higher COD:BOD$_5$ ratios and this was because this source was not only wastewater from a slaughterhouse like the rest but also wastewater from a facility processing and cooking the resulting meat.

Table 2.1.1. Slaughterhouse (poultry) wastewater characteristics.

| Parameters/Cases | Case-1 | Case-2 | Case-3 | Case-4 | Case-5 | Case-6 |
|---|---|---|---|---|---|---|
| $Q_{avg}$, m$^3$ h$^{-1}$ | 24 | 9 | 40 | 66 | 18 | 16 |
| $Q_{pk}$, m$^3$ h$^{-1}$ | 45 | 15 | 70 | 85 | 20 | — |
| COD, mg L$^{-1}$ | 2970 | 2700 | 2000–4000 | 5200 | 2000–2500 | 2300 |
| BOD$_5$, mg L$^{-1}$ | 1480 | 1100 | 1500–3000 | 2500 | 500–750 | 1200 |
| COD:BOD | 2.0:1 | 2.5:1 | 1.3:1 | 2.1:1 | 3.3:1–4.0:1 | 1.9:1 |
| TSS, mg L$^{-1}$ | 950 | 800 | 1000 | 1800 | 1000 | 1000 |
| VSS, mg L$^{-1}$ | 320 | 300 | — | — | — | 400 |
| O&G, mg L$^{-1}$ | 80 | 100 | 200 | 1100 | 150–250 | 150 |
| pH | 6.0–7.5 | 6.0–8.0 | 6.5–7.5 | 6.0–8.0 | 6.0–8.5 | 6.0–7.5 |
| Amm-N, mg L$^{-1}$ | 50 | 40 | 120 | — | 10–190 | 60–70 |
| TKN, mg L$^{-1}$ | 200 | 170 | 200 | 310 | 15–300 | 200–250 |
| Temp, °C | 26–30 | — | 26–34 | 26–34 | 26–34 | 26–35 |

Note: Where two values have been provided — these represent the minimum and maximum composite daily average values noted for a particular parameter over a monitoring period. Single values are the average values of composite daily samples. A "—" means information for that parameter is not available.

Table 2.1.2. Tobacco processing wastewater characteristics.

| Parameters | Values |
| --- | --- |
| $Q_{avg}$, m$^3$ d$^{-1}$ | 150 |
| No. of shifts, d$^{-1}$ | 1 × 8 h shift |
| COD, mg L$^{-1}$ | 4500–11800 |
| BOD$_5$, mg L$^{-1}$ | 760–4200 |
| SS, mg L$^{-1}$ | 140–600 |
| O&G, mg L$^{-1}$ | 10–40 |
| pH | 4.0–5.5 |

Agro-industrial wastewaters need not always have such low COD:BOD$_5$ ratios. For example tobacco processing wastewater (Table 2.1.2) can have a COD:BOD$_5$ ratio of about 6:1. This is a strongly colored (brown coloration arising from the tobacco leaves) wastewater which can be difficult to treat to meet COD discharge limits because residual organics following biological treatment are resistant to further biological treatment.

Examples of more extreme COD:BOD$_5$ ratios may be found in some chemical wastewaters such those arising from dyestuff manufacturing. An example had a wastewater COD of 4400 mg L$^{-1}$ but a BOD$_5$ of only 55 mg L$^{-1}$. The resulting COD:BOD$_5$ ratio was therefore 80:1 which meant biological treatment would unlikely to be successful in removing sufficient quantities of the organics so as to meet the discharge limits (i.e. high effluent COD).

It is important to realize a low COD:BOD$_5$ ratio suggests biological treatment may be successful but does not necessarily mean biological treatment will be successful. A wastewater would have other properties which may be no less important to the success (or failure) of biological treatment. A number of these properties are explored in the following sections.

## 2.2. Strength

Industrial wastewaters often have organic strengths which are very much higher than those encountered in sewage. Agro-industrial wastewaters are among those which may have very high organic strengths. Chapter 10 explores such a strong wastewater, palm oil mill effluent (POME). Table 2.2.1 provides additional examples drawn from tapioca starch, sugar milling and coconut cream extraction industries.

Such wastewaters may benefit from anaerobic pretreatment ahead of the aerobic treatment stage so that organic strength can be reduced and hence reducing

Table 2.2.1. Characteristics of tapioca starch extraction, sugar milling and coconut cream extraction wastewaters.

| Parameters/Industry | Starch extraction | Sugar milling | Coconut cream |
|---|---|---|---|
| $Q_{avg}$, m$^3$ d$^{-1}$ | 3.6 over 8 h | 120 over 20 h | 112 over 24 h |
| BOD$_5$, mg L$^{-1}$ | 2700 | 25000 | 8900 |
| COD, mg L$^{-1}$ | 41000 | 50000 | 12900 |
| pH | 6–7 | — | 4–5 |
| TSS, mg L$^{-1}$ | 23000 | — | 2900 |
| Zn, mg L$^{-1}$ | 25 | — | — |
| O&G, mg L$^{-1}$ | 15 | — | 1560 |
| S$^{-1}$, mg L$^{-1}$ | 0.2 | — | — |
| Cr(total), mg L$^{-1}$ | 0.2 | — | — |
| B, mg L$^{-1}$ | 2 | — | — |
| Phenol, mg L$^{-1}$ | 2 | — | — |
| Temp, °C | — | — | 25–28 |
| TDS, mg L$^{-1}$ | — | 100 000 | — |

the aeration and the consequential energy requirements. Typically the biological processes address the dissolved and colloidal organic components in a wastewater since the particulate component can be easily addressed using physico-removal methods. For example the coconut cream extraction wastewater could have benefited from this approach and certainly the starch extraction wastewater with 23000 mg L$^{-1}$ SS can be treated with a fine screen initially and the 41000 mg L$^{-1}$ COD would then have been very substantially reduced.

However, there are strong wastewaters with SS which may not respond in a similar manner. Table 2.2.2 provides details on distillery wastewater. The wide range in wastewater organic strengths reflect the varying degrees of "dilution" brought about by the merging of various wastewater streams at a distillery. Distilleries can have two major wastewater streams — the fermentation and wash streams. The fermentation stream is usually very strong and would have characteristics somewhat similar to Distillery Case-2 except that the SS which would have been substantially higher. It would have been ineffective to attempt to remove this SS with screens since the material can penetrate even fine screens and can easily blind such screens because of its stickiness.

## 2.3. Volumes

It can be a common misconception that industrial wastewater treatment plants handle volumes which are smaller than sewage flows. While this may be so

Table 2.2.2. Distillery wastewater characteristics.

| Parameters/Cases | Case-1 | Case-2 | Case-3 | Case-4 | Case-5 | Case-6 |
|---|---|---|---|---|---|---|
| Wastewater flow, $m^3\,d^{-1}$ | 42 over 8 h | 60 over 8 h | 1000 over 24 h | 60 over 8 h | 1225 over 20 h | 221 over 24 h |
| $BOD_5$, $mg\,L^{-1}$ | 4000 | 59000–120000 | 3200 | 4100 | 1000 | 15000 |
| COD, $mg\,L^{-1}$ | 6000 | 100000–150000 | 5350 | 9600 | 3000 | 18000 |
| TSS, $mg\,L^{-1}$ | 3500 | 1000–2000 | 900 | — | 180 | 4030 |
| pH | 3–5 | 3.5–4.0 | 4–7 | — | 6.2–7.2 | 6–9 |
| Amm-N, $mg\,L^{-1}$ | — | 1200 | — | — | 1.5 | — |
| TKN, $mg\,L^{-1}$ | — | — | — | — | — | — |
| TP, $mg\,L^{-1}$ | — | — | — | — | — | — |
| TDS, $mg\,L^{-1}$ | 9000 | — | — | 4800 | — | — |
| Temp, °C | 95 | — | — | — | 35 | 105 |
| Feedstock | kaoliang | molasses | rice | rice | mixed grains | — |

when compared with sewage flows received by sewage treatment plants serving metropolitan areas, not all sewage treatment plants serve large communities and not all industrial wastewater flows are small. The range of industrial wastewater volumes to be treated can be very large, not only from one industry to the next but also from factory to factory within an industry. Table 2.2.1 shows an example with only $3.6\,m^3\,d^{-1}$ (starch extraction wastewater) but there are industrial wastewaters, such as those from paper mills (Table 2.3.1) and breweries (Table 2.3.2), with very large volumes. Paper mills are probably among the largest in terms of volumetric loads. For example, Paper Industry Case-1 is equivalent to sewage flows arising from 160000 equivalent population in terms of hydraulic load and 1.7 million equivalent population in terms of BOD load if it had been a sewage flow.

Breweries, although not generating wastewater flows as large as the paper mills, are typically also associated with the larger flows. A major contributor to this large flow of wastewater is the bottling line in the brewery. This is because glass bottles used are returnable and the returned bottles are washed before they can be used to bottle beer again.

The soft drinks industry (Table 2.3.3) has features similar to the breweries in that the bottling lines (where present) also contribute substantially to the wastewater flow. These bottling lines also depend largely on reusable glass bottles which have to be washed before being used again. The bottle washing process (and the dumping of rejected products) can be a cause for the differences in organic strength at the various bottling plants. Apart from the organic components in the wastewater, these washing lines also give rise to debris such as broken glass, bits

Table 2.3.1. Paper industry wastewater characteristics.

| Parameters/Cases | Case-1 | Case-2 | Case-3 | Case-4 |
|---|---|---|---|---|
| $Q_{avg}$, m$^3$ d$^{-1}$ | 27240 over 24 h | 11000 over 24 h | 11000 over 24 h | 4 over 8 h |
| $Q_{max}$, m$^3$ d$^{-1}$ | 36320 over 24 h | 15000 over 24 h | 13800 over 24 h | — |
| BOD$_5$, mg L$^{-1}$ | 2540 | 1950 | 1550 | 850 |
| COD, mg L$^{-1}$ | 5080 | 3500 | 2770 | 6660 |
| TSS, mg L$^{-1}$ | 1600 | 500 | 200 | 490 |
| pH | 5–9 | 7–9 | 7–9 | 8.1 |
| O&G, mg L$^{-1}$ | — | 20 | 10 | 40 |
| TN, mg L$^{-1}$ | — | — | — | — |
| TP, mg L$^{-1}$ | — | — | — | — |
| TDS, mg L$^{-1}$ | — | 1000 | 800 | — |
| Temp, °C | 50–80 | 45–55 | 40–60 | — |
| Phenol, mg L$^{-1}$ | — | — | — | 13 |
| Cu, mg L$^{-1}$ | — | — | — | 8 |
| Mn, mg L$^{-1}$ | — | — | 2 | — |
| Pb, mg L$^{-1}$ | — | — | — | 4 |
| Fe, mg L$^{-1}$ | 5 | — | 5 | — |
| Product | recycled paper | newsprint | recycled paper | cartons |

Table 2.3.2. Brewery wastewater characteristics.

| Parameters/Cases | Case-1 | Case-2 | Case-3 |
|---|---|---|---|
| $Q_{avg}$, m$^3$ d$^{-1}$ | 2500 over 24 h | 800 over 24 h | 700 over 24 h |
| $Q_{max}$, m$^3$ d$^{-1}$ | 4320 over 24 h | 1600 over 24 h | — |
| BOD$_5$, mg L$^{-1}$ | 800–1600 | 600–1500 | 1650 |
| COD, mg L$^{-1}$ | 1250–2550 | 1700–3600 | 2800 |
| TKN, mg L$^{-1}$ | 25–35 | — | — |
| PO-P$_4$, mg L$^{-1}$ | 20–30 | — | — |
| TSS, mg L$^{-1}$ | 150–500 | 270 | 400 |
| pH | — | 4–12 | 6.5–7.5 |
| Temp, °C | 18–40 | 35 | — |

of paper labels, and drinking straws. The inclusion of screens to protect downstream mechanical equipment is therefore an important requirement.

Although six soft drinks cases have been provided as examples, it should be noted that they are not bottling the same product (but all bottle carbonated drinks). The differences in product formulation have also contributed to the differences in wastewater characteristics. A significant impact at a bottling plant would be the number of products produced on a campaign manufacturing basis using the

Table 2.3.3. Soft drinks wastewater characteristics.

| Parameters/Cases | Case-1 | Case-2 | Case-3 | Case-4 | Case-5 | Case-6 |
|---|---|---|---|---|---|---|
| $Q_{avg}$, m$^3$ d$^{-1}$ | 1680 over 24 h | 2500 over 20 h | 400 over 8 h | 720 over 20 h | 1675 over 24 h | 800 over 16 h |
| BOD$_5$, mg L$^{-1}$ | 600 | 1500 | 1500–2000 | 1000 | 800 | 800 |
| COD, mg L$^{-1}$ | 1440 | 3000 | 2500–3000 | — | 2240 | 1410 |
| TSS, mg L$^{-1}$ | 45 | — | 100–300 | 150 | 510 | 460 |
| TN, mg L$^{-1}$ | 3 | — | — | — | — | — |
| TP, mg L$^{-1}$ | — | — | — | — | — | — |
| O&G, mg L$^{-1}$ | 80 | 10–15 | 50–60 | 30 | 10–20 | 10–25 |
| pH | 5.5–10.5 | 3–11 | 2–5 | — | 8.7–9.4 | 8.5–9.1 |
| Detergent, mg L$^{-1}$ | 35 | — | — | — | — | — |
| Fe, mg L$^{-1}$ | 1–6 | — | — | — | — | — |
| Temp, °C | 35 | 35 | 35 | 35 | 35 | 35 |

same bottling lines. This would require shutting down and cleaning the product blending tanks and bottling lines at the conclusion of a particular manufacturing episode resulting in the discharge of stronger wastewater than that experienced at a single product (per bottling line if not per factory) manufacturing premises.

To allow first estimations to be made of wastewater volumes which need to be addressed (especially for situations where a factory has not entered production yet), unit wastewater generation rates shown in Table 2.3.4 can be helpful. It should, however, be noted there are wide variations within any group and this can be due to the different specific products made and processes used within a generic group (e.g. fish processing can include freezing, frying as in fish fingers, and boiling as in shrimp processing). Even when factories are manufacturing the same products and using the same processes, differing housekeeping practices can result in very different unit wastewater generation rates. As such the figures provided should only be used as crude guides and with extreme caution. Pollutant loads, such as BOD and COD loads, may be developed using the figures in Table 2.3.4 with data shown in the other tables in this chapter.

## 2.4. Variations

The study of wastewater characteristics provided in the preceding tables would already show that the wastewaters generated by different factories vary even within the same industry group. This is so for every parameter indicated and particularly so in the case of the volumes of wastewater discharged. In part the variation would have been the result of different quantities of materials processed

Table 2.3.4. Wastewater generation rates for various industries.

| Industry | Unit wastewater generation rate | Additional information |
|---|---|---|
| Soft drinks | 32.4 m$^3$ 1000 bottles$^{-1}$ | Returnable glass bottles |
| Fish processing | 5–15 m$^3$ 1000 kg$^{-1}$ product | Includes frozen and cooked products |
| Fruit & vegetables processing | 0.9–2.0 m$^3$ 1000 kg$^{-1}$ processed material | Includes fruits such as pineapples |
| Canned milk | 2.8 m$^3$ 1000 kg$^{-1}$ product | Sweetened condensed milk |
| Pasteurized milk | 1.8 m$^3$ 1000 L$^{-1}$ product | Usually packed in paper cartons |
| Yoghurt | 5 m$^3$ 1000 kg$^{-1}$ product | — |
| Industrial kitchen | 9.6–12.8 m$^3$ 1000 meals$^{-1}$ | Includes flight kitchens |
| Poultry slaughterhouse | 8.9–20.6 m$^3$ 1000 birds$^{-1}$ | Very largely chickens |
| Papermills | 12–30 m$^3$ 1000 kg$^{-1}$ product | Large component of recycled paper |
| Winery | 2.3 m$^3$ 1000 kg$^{-1}$ product | Grain based |
| Industrial alcohol | 0.1 m$^3$ 1000 kg$^{-1}$ product | Molasses based |
| Sugar milling | 1.5–3.0 m$^3$ 1000 kg$^{-1}$ cane | Sugar cane |
| Pig slaughterhouse | 0.6 m$^3$ animal$^{-1}$ | Serving nearby community |
| Pig farm | 20–45 m$^3$ 1000 spp$^{-1}$ d$^{-1}$ | Washed and not scrapped pens |
| Palm oil milling | 2–3 m$^3$ 1000 kg$^{-1}$ oil extracted | — |
| Palm oil refining | 0.2 m$^3$ 1000 kg$^{-1}$ oil refined | Physical refining |
| Palm oil refining | 1.2 m$^3$ 1000 kg$^{-1}$ oil refined | Chemical refining |

at different locations but even in terms of unit quantity of materials processed there are still variations and this is due to differences in housekeeping practices therein.

Perhaps of greater importance to the designer and operator of a particular plant would be the variations which occur at a given site. For example the preceding tables have shown a $Q_{avg}$ with a discharge period following, e.g., 42 m$^3$ d$^{-1}$ over 8 h. If the discharge period is less than 24 h per day, then it must follow the wastewater is not discharged continuously throughout the day. An 8 h discharge period would have meant the wastewater treatment plant receives no wastewater for 16 h per day unless the holding capacity has been provided so that the wastewater flow can be redistributed over 24 h to ensure continuous flow conditions are met at the treatment plant. It is important to bear in mind that wastewater is typically only produced when a factory is in operation and wastewater flow data provided in terms of m³ d⁻¹ can be misleading.

Table 2.1.1 shows another wastewater flow phenomenon — that there are periods of low and high flows during a day's operation and peak flow — $Q_{pk}$, can be very substantially higher than the average flow, $Q_{avg}$. While $Q_{pk}$ conditions may not last long, such short period high flows or surges can easily upset the unit treatment processes. Surges can be caused by batch discharges (or dumping) which is particularly common at the end of a shift/working day or at the end of a manufacturing campaign.

Although the data presented in Table 2.4.1 may not immediately suggest batch discharges, especially the noodle case which had 30 m³ d⁻¹ over a 24 h period, in both cases, because of the nature of the manufacturing processes used, the discharges had to a very large extent occurred as batch discharges at the end of each shift. This meant that the noodle case had three batch discharges while the vermicelli case had one. A batch discharge with the consequent surge would represent an extreme flow variation situation.

Even in the absence of batch discharges, flows can show wide fluctuations over a day's operation at a factory. This can be due to initiation of certain processes which generate larger volumes of wastewater and subsequently the ending of such processes as activity moves to the next phase of processing. Industrial kitchens (Table 2.4.2) can behave in this manner as activity shifts from preparation of raw materials to cooking and finally packing/serving. Peak flows in such cases may last for a few hours in the working day and sometimes may even be over in less than an hour.

Table 2.4.1. Noodles/Vermicelli manufacturing wastewater characteristics.

| Parameter/Cases | Noodle | Vermicelli |
|---|---|---|
| $Q_{avg}$, m³ d⁻¹ | 30 over 24 h | 35 over 8 h |
| BOD$_5$, mg L⁻¹ | 410 | 1050 |
| COD, mg L⁻¹ | 1000 | 2000 |
| pH | 4–10 | 7–8 |
| TN, mg L⁻¹ | — | — |
| TP, mg L⁻¹ | — | — |
| TSS, mg L⁻¹ | — | 200 |
| TDS, mg L⁻¹ | — | 1000 |
| O&G, mg L⁻¹ | 300–800 | 20 |
| Temp, °C | 25–30 | 26–30 |

Table 2.4.2. Industrial kitchen wastewater characteristics.

| Parameters/Cases | Case-1 | Case-2 | Case-3 | Case-4 | Case-5 |
|---|---|---|---|---|---|
| $Q_{pk}$, m$^3$ h$^{-1}$ | 13 | 21 | 36 | 5 | 525 |
| $Q_{avg}$, m$^3$ d$^{-1}$ | 128 over 16 h | 40 over 6 h | 520 over 24 h | 10 over 8 h | 3645 over 24 h |
| BOD$_5$, mg L$^{-1}$ | 600–800 | 600 | 300–690 | 600 | 500 |
| COD, mg L$^{-1}$ | — | 1400 | 770–1550 | — | 1000 |
| TSS, mg L$^{-1}$ | 200–600 | 400 | 220–580 | 500 | 500 |
| O&G, mg L$^{-1}$ | 100–400 | — | 50–190 | 20 | 350 |
| pH | — | 6.5–8.5 | 6.2–8.9 | — | — |
| Food product | airline | canteen | fastfood | bakery | airline |

Table 2.4.3. Seasonal wastewater variations at a vegetable processing plant.

| Parameters/Periods | Period-1 | Period-2 | Period-3 |
|---|---|---|---|
| $Q_{avg}$, m$^3$ d$^{-1}$ | 550 over 24 h | 350 over 24 h | 400 over 24 h |
| BOD$_5$, mg L$^{-1}$ | 850–1800 | 170–340 | 480–820 |
| TSS, mg L$^{-1}$ | 270–350 | 80–170 | 200–890 |
| TN, mg L$^{-1}$ | 90–170 | 2–20 | 50–190 |
| TP, mg L$^{-1}$ | 10–20 | 1–2 | 20–30 |
| Raw material | peas | beans | potatoes |

While the preceding discussion had focused on the short term variations, a factory can exhibit longer term or seasonal variations and these may be tied to manufacturing campaigns as discussed earlier. Table 2.4.3 shows an example which has three campaigns in a year and each dealt with a different product. In this instance, the change in raw material handled arose out of the different seasonal harvests encountered. The impact of such seasonal harvests on the wastewater treatment plant is substantial as it would not only have to deal with a high flow period which is 1.6 times higher than the low flow period but also daily BOD loads which can be 8 times higher.

The example provided in Table 2.4.3 shows seasonal changes in the raw material handled and consequently changes in the wastewater flow and other characteristics. There can also be seasonal variations which are caused not by campaign manufacturing or changes in the material harvested but by the necessity to process more of the same raw material as the latter's production or harvest peaks. Seafood processing wastewater Case-1 and Case-2 (Table 2.4.4) are examples of such agro-industrial activity. These processing plants are located at ports, receive the catch daily and freeze it. In Case-2 the high season flow is three times higher than the

Table 2.4.4. Seafood processing wastewater.

| Parameters/Cases | Case-1 | Case-2 | Case-3 | Case-4 |
| --- | --- | --- | --- | --- |
| High season $Q_{avg}$, m$^3$ d$^{-1}$ | 200 | 1200 | 135 | 580 |
| Low season $Q_{avg}$, m$^3$ d$^{-1}$ | 150 | 400 | 135 | 580 |
| BOD$_5$, mg L$^{-1}$ | 750 | 3000 | 400 | 4900 |
| COD, mg L$^{-1}$ | 1440 | 4200 | 2000 | — |
| TSS, mg L$^{-1}$ | 350 | 1500 | 1000 | 1130 |
| pH | ~6 | 6.6–7.1 | 6–8 | — |
| TDS, mg L$^{-1}$ | — | 10000 | — | — |
| TN, mg L$^{-1}$ | 25 | — | 90 | 405 |
| TP, mg L$^{-1}$ | 5 | — | — | 95 |
| O&G, mg L$^{-1}$ | — | — | 50 | — |
| Temp, °C | 18–25 | 14–40 | — | — |
| Raw materials | fish | fish, shrimp | tuna | fish, shrimp, tuna |
| Process activity | freezing | freezing | canning | canning |

low season flow. Case-3 and -4 differ because these are downstream processors which cook the seafood and the latter is then canned.

Some seasonal variations may not necessarily be due to peaks in harvest but may be due to peaks in demand. Food related industries can be affected by this, particularly during the period leading to the festive season. In Asia these are the months of October to January. For example, poultry slaughterhouse Case-3 in Table 2.1.1 has a wastewater flow of 550 m$^3$ d$^{-1}$ for 10 months of the year but over 2 months leading to the festive period its flow can increase to 1000 m$^3$ d$^{-1}$. Wastewater quality in such instances need not necessarily change but quantity can change substantially.

## 2.5. Special Characteristics

Industrial wastewaters may have certain characteristics, the effect of which may not be apparent from the sort of wastewater data usually provided. These may, however, have significant adverse impact on the equipment or unit process performance, and aesthetics of a wastewater treatment plant. This section explores a few of these characteristics.

For example, if the COD:BOD ratios of dairy product wastewaters (Table 2.5.1) are considered, the conclusion would be such wastewaters are likely to be easily treated with biological systems. The treatment plant design may then focus on the O&G and BOD strength. Most wastewater treatment plants for milk related wastewaters include O&G removal devices such as oil traps and DAFs. The

Table 2.5.1. Dairy product wastewater characteristics.

| Parameters/Cases | Case-1 | Case-2 | Case-3 | Case-4 | Case-5 | Case-6 |
|---|---|---|---|---|---|---|
| $Q_{avg}$, m$^3$ d$^{-1}$ | 750 over 24 h | 120 over 24 h | 800 over 16 h | 120 over 24 h | 50 over 8 h | 100 over 8 h |
| $Q_{pk}$, m$^3$ h$^{-1}$ | 75 | 40 | 70 | — | — | — |
| BOD$_5$, mg L$^{-1}$ | 1800 | 3400 | 480 | 3000 | 1230 | 940 |
| COD, mg L$^{-1}$ | 3600 | 4300 | 920 | — | 1970 | 1240 |
| TSS, mg L$^{-1}$ | 1000 | 2000 | 120 | 1500 | 440 | 360 |
| O&G, mg L$^{-1}$ | 150 | 1800 | 250 | 2500 | 115 | 40 |
| pH | 3–12 | 6.0–7.5 | 6–8 | 4–11 | 4.3–10.0 | 4.8–6.8 |
| TN, mg L$^{-1}$ | — | 310 | 85 | 260 | 60 | 10 |
| TP, mg L$^{-1}$ | — | — | 1 | — | 40 | — |
| Temp, °C | 26–40 | 26–32 | 30–40 | — | — | — |
| Product | milk based snacks & ice cream | ice cream | condensed milk | ice cream & yoghurt | re-constituted milk | fresh milk |

relatively high O&G content in these wastewaters may be due to the inclusion of vegetable oils in the products (to augment milk fats). Oil traps have been effective at removal of the free O&G but can become sources of strong odor as would any part of the plant which is not regularly cleaned. The use of DAFs can help alleviate this odor problem. Good housekeeping at the treatment plant is, nevertheless, always important. This is because the biological degradation of milk related substances under non-aerobic conditions results in odorous organic compounds. Notwithstanding the organic strength of the wastewater, the selection of an anaerobic biological process for organic reduction prior to aerobic treatment would probably not be an appropriate strategy.

It is important to bear in mind the TSS parameter can be due to many different types of particulate material. A concern which can possibly be associated with some of these particulates is their abrasive properties. While the TSS associated with a dairy wastewater arising from cowsheds may well immediately suggest grit and hence suggest wear on pumps and valves, this may be less obvious in the case of the coffee and sauce industries identified in Table 2.5.2. In the case of coffee (Table 2.5.2 Cases-1 to -3), coffee bean fines contributed substantially to the TSS. These fines have been found abrasive on mechanical equipment. In Cases-4 and -5 (Table 2.5.2), the abrasive component in the TSS which damaged the pumps turned out to be the chili seeds when chili was processed into sauces.

Foaming can be a particularly difficult condition to address at a biological treatment plant's aeration vessels. While there are instances where foaming can be due to the biological process responding to organic loading conditions or certain

Table 2.5.2. Coffee processing and sauce making wastewater characteristics.

| Parameters/Cases | Case-1 | Case-2 | Case-3 | Case-4 | Case-5 |
|---|---|---|---|---|---|
| $Q_{avg}$, m$^3$ d$^{-1}$ | 140 over 24 h | 400 over 24 h | 76 over 16 h | 300 over 24 h | 50 over 8 h |
| $Q_{pk}$, m$^3$ h$^{-1}$ | 7 | 15 | — | 40 | 9 |
| BOD$_5$, mg L$^{-1}$ | 8000–9000 | 1500–2000 | 2660 | 5000 | 800–1480 |
| COD, mg L$^{-1}$ | 11000–12000 | 3000–4000 | 4800 | 10000 | 1800–2880 |
| TSS, mg L$^{-1}$ | 5000–5100 | 500–600 | 1000 | 800 | 130–170 |
| O&G, mg L$^{-1}$ | 100–200 | — | 20–170 | — | — |
| TN, mg L$^{-1}$ | — | — | — | 15 | — |
| TP, mg L$^{-1}$ | — | — | — | 2 | — |
| pH | — | 7.0–7.4 | 4.0–6.5 | 4.5 | 3.0–6.5 |
| Temp, °C | 44–50 | — | 36–42 | 30–45 | 30–42 |
| Product | instant coffee | instant coffee mix with milk and sugar | decafinated coffee | chili & sayo sauce | chili & tomato sauce |

constituents in the wastewater, there are also instances when the wastewater had components within it which can cause foaming even without interaction with the biological process. Detergents are a key component in this latter group. In industrial wastewaters, detergents can appear very frequently because they can be used in cleaning operations at the manufacturing facility. The problem becomes tougher when a facility is manufacturing products which include detergents in their formulations. Examples of these are Case-1, -2, -5 and -6 in Table 2.5.3 which include shampoos or soaps in their list of products. While Case-3 and -4 did not include detergents related products in their list of products, foaming was also observed when these wastewaters were treated.

Aside from the foaming, this group of wastewaters can also be very variable in terms of the specific components they contain if these are tracked over time. This is a consequence of the relatively large numbers of chemicals they use and the campaign nature of their manufacturing activities. For example Case-3 had a minimum of 180 entries on its list of chemicals brought into the factory at any point in time.

Upon careful examination of the characteristics of some industrial wastewaters (and especially specific compounds therein), it is possible to identify components which may require special attention as these may adversely affect the biological treatment process. For example the manufacture of monosodium glutamate, which is used as a flavor enhancer in food preparation, generates a wastewater stream with high concentrations of BOD$_5$ (24000–32200 mg L$^{-1}$) and a COD:BOD ratio of about 2.5. While this may suggest amenability to biological treatment, consideration has to be given to the wastewater's ammonia-N (3200–5000 mg L$^{-1}$)

Table 2.5.3. Personal care and pharmaceutical products wastewater characteristics.

| Parameters Cases | Case-1 | Case-2 | Case-3 | Case-4 | Case-5 | Case-6 |
|---|---|---|---|---|---|---|
| $Q_{avg}$, m³ d⁻¹ | 180 over 8 h | 40 over 10 h | 250 over 24 h | 1000 over 24 h | 130 over 8 h | 100 over 16 h |
| $Q_{pk}$, m³ h⁻¹ | — | — | 40 | — | — | — |
| BOD$_5$, mg L⁻¹ | 2000–3000 | 500–800 | 100–1020 | 4000 | 8200–12400 | 250–400 |
| COD, mg L⁻¹ | 6500–8500 | 2000–3400 | 150–1820 | 8500 | 13400–18500 | 600–800 |
| TSS, mg L⁻¹ | Negligible | 30–40 | 300 | 1500 | 600 | 100–200 |
| O&G, mg L⁻¹ | 100–150 | 400 | — | 500 | 4000–6300 | 25–40 |
| pH | 4–6 | 6.0–7.3 | 6–7 | — | — | 3.5–7.5 |
| TN, mg L⁻¹ | 100–125 | — | 15–30 | 130 | — | — |
| TP, mg L⁻¹ | — | — | 0–3 | 30 | — | — |
| Sulphate, mg L⁻¹ | 100–150 | — | — | — | — | — |
| Sulphide, mg L⁻¹ | — | — | 20 | — | — | — |
| Temp, °C | 30–35 | — | — | — | — | — |
| Product | Cough drops & shampoo | Personal care — incl' shampoo | Pharma' incl' antibiobitics & vitamins | Pharma' — nutritionals | Soaps | Personal care — incl' shampoo |

and sulphate (25000–40000 mg L⁻¹) contents. A similar difficulty can be encountered when handling rubber serum wastewater which has lower but still substantial amounts of ammonia-N (210 mg L⁻¹) and sulphate (4500 mg L⁻¹). Aside from the preceding, metals may also be encountered. Zinc can be frequently encountered in rubber related wastewaters. Rubber thread manufacturing generates large volumes of wastewater with relatively high BOD$_5$ values (4000 mg L⁻¹) of which acetic acid would be the main component. The latter is an easily biodegradable component. This wastewater is, however, difficult to treat because of the presence of zinc (250 mg L⁻¹). An example from the food industry is aspartame which is used as an artificial sweetener in the soft drinks industry. This component has been noted to cause some difficulty in terms of process stability during biological treatment.

Food industry wastewaters can be difficult to treat because of the slug discharges of disinfectants whenever a plant shutdown and clean-up takes place. This can occur as frequently as at the end of each shift. Examples of disinfectants which may be encountered include paracetic acid, hydrogen peroxide, chlorine and sodium hypochlorite. The slug entry of such compounds into the biological process basin would likely destroy the microbial culture therein and this would in effect have ended the plant's ability to treat the wastewater.

Treating chemical industry wastewaters is frequently known to be difficult. While this may be due to the presence of specific components which are inhibitory or resistant to biological degradation (as in the dyestuff wastewater discussed in Sec. 2.1), the difficulties experienced may also be due to the presence of large quantities of very easily degradable organics. For example the organic component in a vinyl acetate wastewater is largely made up of acetic acid. The presence of such an easily degradable component can lead to bulking sludge in the activated sludge (or its equivalent) process. Bulking sludge in turn leads to poorer settled effluent quality, difficulties in adequate sludge return (and hence eventual process failure), and higher moisture content in the dewatered sludge.

## 2.6. The Manufacturing Process

Some knowledge of the manufacturing process can be helpful in understanding wastewater characteristics. For example a factory typically collects all its wastewater streams before channeling these collectively to the wastewater treatment plant in a single pipe or drain. An understanding of the streams contributing to the combined wastewater stream may help reduce the volume which requires treatment. Case-2 in Table 2.5.3 has a total wastewater flow of $40 \, m^3 \, d^{-1}$ but $23 \, m^3 \, d^{-1}$ of this is an overflow from the cooling processes and this latter stream would not require treatment to meet the discharge limits. Removing this stream would reduce treatment of the wastewater flow to $17 \, m^3 \, d^{-1}$, which is a very substantial reduction.

In other instances, knowledge of the manufacturing process sequence may allow a particular stream to be intercepted for pretreatment before it is allowed to join the rest of the wastewater streams for further treatment. Table 2.2.1 highlighted the characteristics of a coconut cream extraction wastewater. The coconut processing sequence is as follows: (1) receiving coconut fruits, (2) shelling, (3) raring, (4) washing the kernel, (5) grinding the kernel, (6) pressing the ground kernel for milk, (7) spray drying the milk, (8) homogenizing the resulting cream and, (9) packaging the coconut cream product. Wastewater streams arise from stages (3), (4), (5), (7) and (9). This is a wastewater with considerable amounts of TSS and much of this comes from equipment washing in stage (5) — grinding. If this stream of wastewater had been intercepted for screeming, then the size of the screen could have been much smaller given the smaller hydraulic load.

In a condom manufacturing plant the sequence of manufacturing activities is as follows: (1) withdrawing latex from storage, (2) compounding, (3) pre-aging, (4) leaching, (5) rinsing, (6) acid cleaning, (7) rinsing, (8) cleaning, (9) latex dipping, (10) powdering, (11) vulcanization, (12) depowdering, (13) pinhole testing and, (14) product packing. Continuous wastewater flows arises from stages (2), (5), (7) and (8) but knowing when the batch discharges from stages (3), (4) and (6) occur can help the designer and operator anticipate the surges in terms of timing, volume, and strength.

Occasionally issues can only become apparent after observing factory operations. To illustrate this, consider the appearance of mineral O&G in wastewater from a soft drinks bottling plant. Mineral O&G obviously would not have been in the drinks formulation and should not appear anywhere in the sequence of activities leading from preparing the drinks to bottling. A day spent beside a bottling line, however, showed the line operator oiling the bottle conveyor belt at regular intervals to ensure smooth movement. Excess oil dripped onto the floor and this was then washed, with other drippings and spills, into the drains leading to the wastewater treatment plant and hence the appearance of substantial quantities of O&G at the treatment plant.

# CHAPTER 3

# THE SEWAGE TREATMENT PLANT EXAMPLE

## 3.1. The STP Treatment Train

Any wastewater treatment plant, with no exception of the sewage treatment plant, is a combination of separate unit processes arranged in a sequence such that each would support the performance of the downstream unit process or processes as wastewater with a particular range of characteristics progresses through the plant. This sequence of unit processes forms the treatment train. At the end of this treatment train, the resulting effluent is expected to meet a specified quality. The amount of treatment, and hence the complexity of the plant, is dependent on the treated effluent quality objectives and the nature of the raw wastewater. Notwithstanding the size and engineering complexity of some of these treatment plants, the unit processes in these plants can be classified into five groups:

  (i) Preliminary treatment;
 (ii) Primary treatment;
(iii) Secondary treatment;
(iv) Tertiary treatment and;
 (v) Sludge treatment.

Readers who are familiar with sewage treatment plants (STPs) would have recognized the sequence of treatment stages described above. Sewage treatment plants typically include Stages 1, 2, 3, and 5 although increasing numbers of plants can now include Stage 4, tertiary treatment, as well.

To provide a frame of reference for the reader as he/she progresses through the remaining chapters, this chapter provides a brief description and discussion of the unit processes in STPs. Subsequent chapters would then draw the reader's attention to the possible differences one may encounter in industrial wastewater treatment plants (IWTPs), as compared to STPs, because of the differences in characteristics between industrial wastewaters and sewage. The treatment train of a STP without tertiary treatment can comprise of the inlet pump sump with its racks or bar screens, grit removal, primary sedimentation, biological treatment process, secondary sedimentation and disinfection before discharge of the treated effluent. In some STPs nowadays, the primary clarifiers may be replaced by mechanically cleaned fine screens. Sludge from the primary clarifier and waste activated sludge from the secondary clarifier would be thickened, stabilized (typically by aerobic means in small plants and anaerobically in large plants), conditioned, dewatered, and the resulting sludge cake disposed off.

## 3.2. Preliminary Treatment

The front boundary limit of a STP is typically the inlet pump station. The incoming sewer discharge the sewage into a pump sump and the inlet pumps therein would lift the sewage up to the level of the headworks in the plant. The incoming sewer draws its sewage from a sewer network designed to collect sewage from all the individual sources located within its catchment. The inlet pump sumps of STPs can be deep with the deeper pump sumps associated with the larger plants. This is because the larger plants serve larger communities and this would have meant more extensive sewer networks and hence larger distances covered. The depth of the pump sump is determined by the necessity for an appropriate hydraulic gradient to ensure, where feasible, gravity flow of sewage in the sewer towards the STP. The reachable depth of the sloping sewer as it traveled from its catchment to the STP would be at its greatest just when it reaches the inlet pump station. On occasions when distances are large and the sewer would have been too deep if it were to run uninterrupted from catchment to STP, sewage pump stations may be inserted at intervals to lift the sewage and then to allow it to flow by gravity to the next pump station before being lifted again.

Preliminary treatment takes place at the headworks. This stage can also include flow measurement, but does not change the quality of the sewage substantially in terms of the typically monitored effluent quality parameters (eg. $BOD_5$). It enhances the performance of downstream processes by removing materials which may interfere with mechanical, chemical, or biological processes. For example the racks and coarse screens used are intended to remove relatively large sized suspended material and such devices typically have screen apertures of 25 mm or larger. These devices may be manually cleaned as in basket screens or automatically cleaned as in mechanically raked bar screens. Material collected on such screens can include rags and plastic bags (Fig. 3.2.1) and these can damage downstream mechanical equipment such as pumps by binding the impellers. The material collected on these racks and screens would be removed regularly to avoid odorous conditions from developing and to prevent blinding of the screens when too much material has collected on it.

Fig. 3.2.1. Example of screenings collected on a manually cleaned rack. Note the gross material which includes pieces of paper and plastic wrapping. These can blind the screen unless regularly removed.

The mechanical equipment in contact with sewage may also suffer from excessive wear caused by the grit present in the latter. Grit is inert inorganic material such as sand particles, eggshells, and metal fragments. Grit removal devices rely on differences in specific gravity between organic and inorganic solids to effect separation. It is important the device does not remove the organic solids but to allow these to continue with the sewage flow to the next unit process. Grit removal devices may look like rectangular channel-like structures or the more compact circular chambers. The channel-like devices are frequently aerated along one side of the channel to assist the separation by creating a rolling motion in the water as it flows through while the circular devices would rely on centrifugal forces as sewage is injected tangentially into the chamber.

Aside from grit, sewage may also contain quantities of oil and grease (O&G). The bulk of this O&G is associated with cooking in the homes and is therefore organic in nature. The mineral oil content can be expected to be low. Excessive O&G combined with particulates may blind downstream screens. The O&G may then continue into the aeration basins and interfere with oxygen transfer in the biological processes there. Excessive quantities of O&G entering these biological reactors may also result in "mud-balling" of the biomass where the latter agglomerate into small ball-like structures.

Process performance may then deteriorate because of diminished contact between the microbial population and substrate. Where it is considered an issue, the O&G is removed with O&G traps. These are often baffled tanks with manual or mechanical skimmers for the removal of the free O&G which has floated to the surface of the water during the time the water spends in the trap and is then retained against the baffle. Like the screenings on the racks and screens, the trapped O&G has also to be regularly removed to avoid formation of odorous conditions.

## 3.3. Primary Treatment

Primary treatment follows the preliminary treatment stage. The purpose of primary treatment is to remove settleable suspended solids (SS) and typically about 60% of these may be so removed with unaided gravity settling. While a small portion of the colloidal and dissolved material may be removed with the SS, this is incidental. Notwithstanding this, 30~40% of the $BOD_5$ in the raw sewage may be removed with the SS. In gravity clarifiers, the relatively quiescent conditions therein would allow the settleable solids to settle to the bottom of the clarifier forming a sludge layer there. To achieve such settling conditions, the surface overflow rates chosen for design and operation of a clarifier usually range from 0.3 to 0.7 mms$^{-1}$. In large clarifiers a scrapper located near the base of the clarifier moves the sludge into a hopper from where it would be pumped to the sludge treatment stage. The settled sewage exits the clarifier by overflowing the outlet weirs. Typically these weirs extend around the periphery of the clarifier. This is to accommodate the weir overflow rate deemed appropriate for a particular design. Where there is such a necessity, the weir length may be extended by supporting the launder on brackets some distance from the wall of the clarifier. Large STPs typically operate either circular or rectangular clarifiers while the smaller ones can use either circular or square clarifiers.

Primary (and secondary) clarification in STPs is typically unaided in terms of coagulant use. Where coagulants have been used, SS and $BOD_5$ removals up to 90% and 70% respectively have been achieved. While the application of coagulants on a large scale in sewage treatment is relatively rare in Asia, it has appeared where there is a requirement to remove phosphorous. The coagulant may then be injected before primary clarification or into the biological aeration vessels.

While primary treatment is usually achieved with gravity clarifiers, rotating and static fine screens have been used sometimes. Such screens typically have screen openings of about 0.8 mm to 2.3 mm. Since fine screens are operated at hydraulic loading rates an order of magnitude higher than those applied on clarifiers, they occupy much less space for equipment installation. If the sewage contains substantial quantities of O&G, then the screen would likely to be located after the O&G trap. This reduces the risk of the O&G combining with fine particulates and blinding the fine screen. Fine screens are not expected to remove as much of the SS and $BOD_5$ as primary clarifiers would. Consequently a STP which has fine screens in place of primary clarifiers would need to have its secondary treatment stage appropriately sized.

Primary clarifiers and screens can be major sources of malodors. Avoiding over-designs especially in clarifiers (resulting in overly long hydraulic retention times and the consequent development of septic conditions) and good housekeeping would help reduce the incidence of such odors. The development of septic conditions in screens is less likely to occur since the passage of sewage through the screen does provide a degree of aeration.

## 3.4. Secondary Treatment

The role of the secondary treatment is to remove the colloidal and dissolved material remaining after the preliminary and primary treatment stages. In sewage treatment, the secondary stage typically includes a biological process. The latter, often an aerobic suspended growth process where the microbial population used to treat the wastewater is suspended in the mixed liquor of the reactor, is housed in an aeration vessel or reactor which has been designed to be complete-mix, plug-flow, or a condition between these two extremes — arbitrary flow (see Sec. 5.3 for discussion on reactor configurations). These reactor variants best suit specific process variants. The latter include the high-rate activated sludge, conventional activated sludge, and extended aeration process. Among the differences between these process variants, two important ones are the hydraulic retention time (HRT) and the cell residence time (CRT). Typically the high-rate activated sludge process has the shortest HRTs and CRTs and these parameters would increase in magnitude towards the extended aeration process. This means, for a given reactor volume, the high-rate activated sludge system processes more sewage than the extended aeration system. The latter makes up for this "inefficiency" by usually being a more stable process and therefore easier to operate. Even within the three process variants identified above, there are further variants. For example the oxidation ditch and aerated lagoons are two variants of the extended aeration process but housed in different reactor designs — plug-flow and arbitrary flow respectively. All these variants have the reactors followed by secondary clarifiers. The latter serves to produce a treated effluent with 50 mg L$^{-1}$ SS or lower and allow the return of biomass (or biosludge) collected in the hoppers of such clarifiers to the reactors so as to maintain an adequate microbial population or mixed liquor suspended solids (MLSS) therein.

While aerobic suspended growth systems are common they are by no means the only types used for sewage treatment. Attached growth systems such as the trickling filter and rotating biological contactor may also be encountered in STPs. These systems have the micro-organisms forming a biofilm on a support medium which is typically a highly porous formed plastic shape with a large surface area to volume ratio. Such biofilms are not submerged in sewage (eg. in the trickling filter) or only intermittently submerged (eg. in the rotating biological contactor). Oxygen for the aerobic process is transferred from the atmosphere into the liquid film which forms on the biofilm.

Figure 3.4.1 shows an activated sludge process variant which combined suspended growth with attached growth. Unlike the trickling filter and rotating biological contactor, the biofilms in such a system are continuously submerged in the reactor's mixed liquor. Since the biofilm support medium is submerged, its

Fig. 3.4.1. Activated sludge plant for sewage treatment where the bioprocess is a combination of suspended and attached growth. The submerged biofilm support modules are located along a line running longitudinally and down the center of the tank.

presence is not immediately obvious. Its presence is, however, suggested by the aeration pattern observable on the water surface. Because the diffusers have been concentrated beneath the support medium, the distribution of air bubbles on the water surface is not even as it would have been in an aeration vessel where the diffusers had been distributed evenly on the base of the vessel. In such processes, the biological reactor would also be followed by a secondary clarifier. This is to allow return of settled biomass to the reactor so as too maintain an adequate population of suspended microbes therein. The clarified effluent overflows the clarifier and can be discharged into a receiving waterbody if it does not require further treatment.

## 3.5. Sludge Treatment

Biomass in excess of the quantity required for maintaining the MLSS concentration in the reactor is removed from the system via the excess sludge line. The waste sludge can be thickened in gravity thickeners and then aerobically or anaerobically digested to reduce solids content and to render the sludge safer in terms of pathogenic organisms (especially if the waste activated sludge had been mixed with primary sludge from the primary clarifiers. Typically anaerobic digestion is used at large STPs while aerobic digestion would be used at those serving small communities. The anaerobic process in STPs is almost always used for treating the solids rather than the liquid stream. Anaerobic digesters at large STPs, aside from reducing the quantity of solids requiring final disposal, also offer the opportunity for recovering energy from the organic solids. The digested sludge is dewatered to reduce moisture and hence volume. Methods used include drying beds, filter presses, and centrifuges. Nowadays drying beds are rarely used at large STPs because of their large space requirements. The resulting sludge cake from the dewatering stage is disposed off as a soil conditioner, at landfills, or incinerated.

## 3.6. An Alternative Plant Configuration

The preceding description has a continuous flow biological treatment stage. If this is substituted with a cyclic biological treatment stage then secondary clarifiers are not required. The plant begins with coarse to medium screens (Fig. 3.6.1) and macerating pumps in the inlet pump sump before the sewage enters one of the chambers in the cyclic sequencing batch reactor (SBR) sub-system. Do note that an equalization tank may be found between the inlet pump sump and the SBR tanks. SBRs may operate with one or more tanks (frequently a single large tank is constructed and this is then divided into chambers). Each chamber receives

Fig. 3.6.1. Mechanical screen (LHS) at the headworks of a STP (first of a pair installed).

sewage in turn and each would operate through a sequence of FILL, REACT, SETTLE, DECANT, and IDLE. The screened sewage thus enters the SBR, is treated, and clarified before discharge. During treatment, air is provided to maintain aerobic conditions. Aeration can be by surface aerators or submerged diffusers (Fig. 3.6.2). Given the absence of primary clarifiers in many of the smaller STPs using the SBR, the aeration capacity would have to accommodate the higher pollutant load reaching the SBR chambers. Where air diffusers are used, air blowers would supply the air. These are typically provided with at least one serving as standby (Fig. 3.6.3). While primary clarifiers may or may not be present, secondary clarifiers are certainly absent. Given the absence of the latter, excess sludge wasting has to be from discharge points located near the base of the SBR chambers while clarified effluent is also decanted from the chamber but via a moving weir or decanter arrangement (Fig. 3.6.4).

## 3.7. Tertiary Treatment

In cases where plants include some form of tertiary treatment, there can be unit processes following the secondary clarifiers or, in the case of the cyclic systems,

Fig. 3.6.2. Air header and circular membrane diffusers in a reactor drawn down for maintenance.

the decant sumps. In sewage treatment, disinfection of the clarified treated effluent has become an increasingly frequent requirement. Either calcium or sodium hypochlorite is often used as the disinfectant with perhaps calcium hypochlorite being preferred at the larger installations because of its lower cost compared to other readily available disinfectants. Calcium hypochlorite is supplied in either

Fig. 3.6.3. Blowers sited in a blowerhouse which also served to attenuate noise (blowers installed but still to be commissioned).

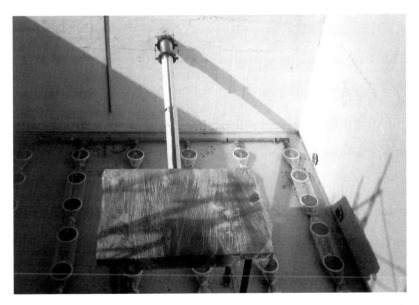

Fig. 3.6.4. Decanter assembly in a cyclic reactor (SBR) with the overflow weir in the decanter head (foreground) (decanter and diffusers installed but still to be commissioned). The decanter assembly is articulated at the point where the decanter head meets the decanter arm and where the arm meets the discharge pipe exiting the reactor's wall.

Fig. 3.7.1. Labyrinthine-type disinfection tank at a STP. The tank has been drawn down for maintenance. The arrangement of the baffles generates eddies which help in the mixing.

powder or liquid form. Should it be in the former form, it would be made into a solution and then injected into labyrinthine-type chambers. The latter provide both mixing and the contact time necessary for the disinfectant to act (Fig. 3.7.1). Where such labyrinthine-type tanks are not used, mixing can be provided by a mechanical stirrer, hydraulic jump downstream of a weir, or inline dosing and mixing of the disinfectant before contact in a tank.

## 3.8. Summary

Mechanical equipment, aside from control at point, would be connected to the central control panel for operation and monitoring of all plant equipment at a single location (Fig. 3.8.1). Figure 3.8.2 shows a package sewage treatment plant based on the cyclic SBR design while Fig. 3.8.3 shows a much larger sewage treatment plant again based on the cyclic SBR design.

In summary a STP would have a preliminary stage comprising screening to remove gross particles, degritting to remove grit, and possibly a grease trap to

## The Sewage Treatment Plant Example

Fig. 3.8.1. Central control panel at a small STP.

Fig. 3.8.2. Package STP based on the SBR for a small community of 2000ep. Note the steel plate construction of the vessels. These can be welded together relatively quickly and mounted on the concrete plinth.

Fig. 3.8.3. Mid-sized STP based on the SBR for 20000ep. This larger STP, compared to the example shown in Fig. 3.8.2, has been constructed using reinforced concrete. The decision between using steel or concrete is often based on the speed of construction required and the cost of the construction materials.

remove oil and grease. This would be followed by the primary stage with its primary clarifiers or medium to fine screens to remove the smaller particles. Secondary treatment would typically be provided by an aerobic biological process such as the activated sludge process. The biological system typically comprises of a vessel or vessels for biological reactions and thereafter secondary clarification to separate the activated sludge from the treated effluent. The biological process is intended to remove the colloidal and dissolved material in the wastewater. Given the discharge limits on ammonia, the biological process is likely to have been designed to nitrify the wastewater resulting in the conversion of ammonia to nitrates. Treated and clarified effluent may then be disinfected before discharge while excess sludge can be digested anaerobically at large plants and aerobically for the smaller plants before dewatering and final disposal. This sludge is largely organic in nature. Where nutrients (nitrogen and phosphorus) removal is required, then the biological process would be designed to include denitrification so that the nitrates formed during bioxidation of the sewage can be converted to nitrogen gas. Phosphorus, if also required to be removed, can be removed by biological accumulation in the biomass and then removed when excess sludge is wasted.

Biological removal of phosphorus can be backed up with chemical precipitation using either lime or alum. This latter option would change the quantity and nature of the sludge generated at the STP and requiring final disposal. Quantities can increase substantially and the sludge then has a high content of lime or aluminum depending on the coagulant used.

# CHAPTER 4

# THE INDUSTRIAL WASTEWATER TREATMENT PLANT — PRELIMINARY UNIT PROCESSES

## 4.1. The IWTP Treatment Train

In industrial wastewater treatment plants (IWTPs) there is a treatment train with unit processes arranged in a manner similar but not necessarily identical to that found in STPs. It is necessary to bear in mind that unit processes present in a STP can all be present in an IWTP treatment train or many may not be present. Unit processes not typically found in STPs may also appear in an IWTP treatment train. Much depends on the industrial wastewater's characteristics and treatment objectives. This can result in significant differences between IWTPs and STPs. The latter, because of the greater similarities in sewage characteristics from location to location, tend to have a more recognizable arrangement of unit processes and plant configuration (as discussed in Chapter 3). Differences in the latter can often come about primarily because of plant size instead of the sewage's characteristics. Because of the differences in industrial wastewater characteristics, as discussed in Chapter 2, this is not so for IWTPs.

## 4.2. Wastewater Collection and Preliminary Treatment

IWTPs, like STPs, also begin with a sump where the inlet pumps are located. These sumps serve to collect the wastewater from the factory before onward transmission to the IWTP. These sumps are, however, very rarely as large as those which may be found in STPs. This is because most IWTPs, but not all, serve a single wastewater source, which is the factory for which it has been constructed for. Since the IWTP is often located close to the source of its wastewater, the inlet (or collection) sump is also rarely very deep (Fig. 4.2.1). This is because wastewater pipes leading from the factory to the IWTP rarely need to be placed in deep trenches at the IWTP end to ensure an adequate slope to facilitate wastewater flow. The incoming wastewater may, in fact, often arrive at the sump by way of surface drains instead of buried pipes. Where drains are used instead of pipes, care

Fig. 4.2.1. Shallow wastewater collection sump at a personal care products factory. Submersible are located in this sump to lift the wastewater to the next unit process in the IWTP. The sheen on the water surface suggests the presence of O&G.

should be exercised to separate rainwater from the roof gutters and surface runoff from the wastewater flow. This is an important consideration at locations where seasonal rainfall can be heavy over relatively short periods of time. The resulting surge of high flows arising from rainwater can easily overwhelm an IWTP in terms of hydraulic capacity leading to, for example, washout of oil and grease from O&G traps and biomass from the bioreactors. Figure 4.2.2 shows a drain leading to an IWTP which has been covered to reduce entry of rainwater runoff.

The drains leading to the collection sump may offer opportunities for the inclusion of preliminary treatment devices such as simple bar screens and O&G traps. Figure 4.2.2 which shows a wastewater drain leading to an IWTP's collection sump also shows a bar screen inserted into it. This has a function similar to the bar screens in STPs. Figure 4.2.3 shows a drain leading to another collection sump. In this instance a simple perforated baffle plate has been mounted in it. This served to remove some of the O&G present in the raw wastewater and is therefore a simple oil trap. The drain may eventually lead to a baffled tank O&G trap as shown in Fig. 4.2.4. Application of such O&G traps early in the treatment train (eg. in the drains leading to the IWTP) is useful for wastewaters such as palm oil refinery effluents where the suspended solids content is relatively low while the O&G

Fig. 4.2.2. Covered drain leading to the collection sump of an IWTP. The (removed) cover had been placed over a coarse screen. The easily removable covers facilitate the removal of screenings from the screen. The drain is shallow because the distance between the factory and IWTP is small.

content can be high. They can be important for enhancing the performance of downstream mechanical elements such as pumps and valves (reducing the risk of clogging), and unit processes such as the dissolved air flotator (DAF) (reducing the O&G load). DAFs may be required if there is a downstream requirement for relatively low residual O&G content — levels which cannot be met by the simple or baffled tank O&G traps. The simple upstream O&G traps do considerably reduce the O&G load on the DAF and this would reduce the size of the DAF and quantities of air and coagulant required therein to achieve the desired performance. Even if such downstream processes are not present, the O&G traps would have helped make housekeeping at the equalization tank easier. The organic O&G recovered from such traps are frequently collected by manufacturers of coarse soaps. For example if it is palm oil, it would be hydrolyzed with hydroxide ions in aqueous solution (ie. saponified) and sodium palmitate ($CH_3(CH_2)_{14} \cdot CO \cdot ONa$), which is soap, is produced.

The possible presence of O&G need not always be indicated by the nature of the industry — as in a palm oil refinery obviously generating wastewater with

Fig. 4.2.3. Baffle plate O&G trap inserted into a surface drain leading to the IWTP at a palm oil refinery. The effectiveness of such a simple device may be seen from the O&G accumulated behind the baffle plate. While not reducing the wastewater's O&G content to the required levels, the trap significantly reduced the O&G load which would otherwise be imposed on the next unit process.

O&G. An example of a less obvious case would be the personal care products factory wastewater. The O&G in this instance was part of the formulation of the products resulting in a "greasy" wastewater. The wastewater shown in Fig. 4.2.1 obviously shows the presence of O&G by way of its surface sheen. Even less

Fig. 4.2.4. The baffled tank O&G trap at a palm oil refinery provides for more quiescent conditions to allow for greater removal of O&G than what is possible with the simple trap shown in Fig. 4.2.3. In traps of this type, the O&G accumulated on the wastewater surface in each chamber is manually removed at intervals.

obvious than the personal care products factory is the case of the soft drinks bottling plant. Figure 4.2.5 shows the mineral oil collected by an O&G trap at such a bottling plant over a week. If this quantity of O&G had entered the downstream bioreactor, it would have affected the dissolution of oxygen into the reactor's mixed liquor during aeration, caused the biomass to form small lumps resembling "mud balls", and compromised treated effluent quality. The reason for the occurrence of this O&G has been discussed in Chapter 2.

## 4.3. Wastewater Equilization

Unlike many STPs, IWTPs would frequently include equalization tanks in their treatment trains. These serve to produce flows, or compositions, or both which are closer to the average values used in the IWTP designs. In addition to this, and bearing in mind that factories are operated on the basis of shifts and if a particular factory is operating on fewer than three 8 h or two 12 h shifts per day, the equalization tank can also have the function of a holding tank so that wastewater can be stored and supplied continuously to a continuous flow IWTP even when

Fig. 4.2.5. Lubricating oil collected from the O&G trap at a soft drinks bottling plant. The lubricating oil came from excess oil applied to the bottle conveyor belt system. This dripped onto the floor and was eventually washed into the collecting drains leading to the IWTP.

the factory has ceased operations and stopped discharging wastewater for the day. Where variable wastewater composition is an issue, the contents of the equalization tank would need to be mixed. Mixing is also important if the wastewater contained settleable material. In the absence of mixing, such material would settle and accumulate in the equalization tank. Mixing can be achieved with mechanical mixers or by aeration. Although the latter is typically performed through coarse air aeration from coarse air diffusers or perforated pipes, some dissolution of oxygen would occur and this is useful if there is a concern that septic conditions may develop as biodegradable substances present in the wastewater degrade over the holding period. Figure 4.3.1 shows an aerated equalization tank which served a lanolin extraction factory. The latter had a process which included washing wool prior to extraction of lanolin from the resulting wash water. The waste wash water contained significant quantities of O&G and very fine particulate material, giving the wastewater a thick brownish appearance. The equalization tank shown is vigorously aerated to reduce settling of the particulate material and had been designed with two chambers to facilitate cleaning in view of a wastewater which can easily foul the tank's fittings, walls, and base as shown.

Fig. 4.3.1. Two-chambered aerated equalization tank at a lanolin extraction IWTP. The RHS chamber was being aerated while the LHS chamber was not at the time this picture was taken. The mixture of O&G and fine particulates make fouling of the equalization tank a recurring and serious maintenance issue. The material on the walls and pipe (foreground of the picture) gives an indication of the fouling. The latter had to be removed regularly to reduce the incidence of odours and slippery work surfaces.

Where factories are not known to have shutdown periods and bypassing a unit process is undesirable, the latter, such as the equalization tank shown in Fig. 4.3.1, may be designed in pairs (for vessels this would more typically be two chambers) to facilitate partial shutdown of a stage in the treatment train so that maintenance can be performed. It is necessary to allow for some redundancy in an IWTP. Without the redundancy, plant maintenance can then only take place when the factory has a shutdown.

Equalization tanks have also served to hold and so cool a warm wastewater stream prior to its treatment. Warm wastewaters are a common occurrence in food processing and canning factories. In general, warm wastewaters occur more frequently than wastewaters with temperatures substantially below ambient. Factories often generate a number of wastewater streams with different temperature characteristics and the equalization tank then serves as a blending tank so that the IWTP may receive a blended wastewater with more consistent thermal characteristics. These different wastewater streams may come from the different manufacturing lines within the factory. The importance of an adequately sized and properly operated equalization tank to overall IWTP performance cannot be overstated. Failing this, IWTP performance is unlikely to be stable and hence the treated effluent may not meet the discharge limits consistently.

## 4.4. Oil & Grease and Particulate Removal

Large quantities of fine particulates and O&G, which have been found in lanolin extraction wastewater, would impose high SS and organic loads on the bioreactors downstream. This would have led to higher oxygen demands. Sizing of such bioreactors, and possibly other types of downstream unit processes, can be reduced if some of the pollutants are removed upstream. As pointed out earlier (Sec. 4.2), DAFs may serve to remove pollutants such as O&G and fine particulates. Fig. 4.4.1 shows the lanolin extraction wastewater first shown in Fig. 4.3.1 after DAF treatment. The improved clarity of the DAF treated wastewater is obvious indicating that the pollutant load had been much lowered, at least in terms of the fine particulates and O&G. To achieve such an improvement in wastewater quality by the DAF often requires the use of coagulants. Among the coagulants used, aluminum (alum — $Al_2(SO_4)_3 \cdot 14H_2O$) and iron salts (ferrous sulphate — $FeSO_4 \cdot 7H_2O$; ferric chloride — $FeCl_3 \cdot 6H_2O$) are common. This is usually because of the relatively low cost and availability of these chemicals. In each case the coagulant reacts with the alkalinity in the wastewater and forms the metal hydroxide as in $Al(OH)_3$ or $Fe(OH)_3$. Since wastewaters may not have sufficient alkalinity to react with the amount of coagulant added, alkalinity has to be supplemented. Alum coagulation is generally effective over a pH range of 5.5 to 8.0 whereas the iron salts can be effective over a wider pH range of 4.8 to 11.0. pH control is an important consideration in coagulation as the solubility of the metal hydroxides increase outside of the optimum conditions determined for each wastewater. Precipitation of the amorphous metal hydroxide is a requirement for most coagulation and clarification processes to work effectively.

The use of coagulants in wastewater treatment such as to assist air flotation is not without issue. The largely metal hydroxide sludge so generated requires disposal at landfills thereby increasing the overall cost of wastewater treatment. Large quantities can be generated when treating strong wastewaters. The example provided in Fig. 4.4.2 is of a DAF which treated wastewater from a milk canning factory located in an urban area. Anaerobic treatment as a means to reduce wastewater organic strength before aerobic treatment was not acceptable because of the potential odours from the former and consequent objections from neighbors. The coagulant assisted DAF was therefore used to remove O&G, and to reduce overall organic strength before the pretreated wastewater was discharged directly

Fig. 4.4.1. Coagulant assisted DAF pretreated lanolin extraction wastewater — Note the substantial improvement in clarity. Significant quantities of the O&G and fine particulates have been removed. Although the wastewater's clarity had improved so markedly, it still needed biological treatment to remove the dissolved organic components before the treated effluent met the discharge limits.

Fig. 4.4.2. Coagulant assisted DAF (top LHS) treating milk canning wastewater. The pretreated wastewater is discharged directly into the bioreactor beneath the DAF for further treatment. The bags on the RHS contain dewatered sludge which was mainly made up of coagulation sludge from the DAF.

into the activated sludge basin sited below the DAF unit. The bags beside the aeration basin show how much dewatered sludge can be accumulated over a month. DAFs do not lend themselves well to intermittent operation as time is required to stabilize the process following each start-up. As such it is desirable to operate DAFs continuously and this would require an appropriately sized equalization tank. Apart from the continuous and constant hydraulic load, the equalization tank would also have "averaged" the wastewater's composition thereby allowing the possibility of lower chemicals consumption. Not over-sizing the equalization tank is, however, an important consideration for the same reason an anaerobic pretreatment stage was thought inappropriate for milk wastewater treatment.

Other unit processes which may be used to remove particulates, especially the coarser and/or denser types, include the primary clarifier and fine screen. As with the DAF, primary clarifiers in industrial wastewater treatment are often preceded by coagulation and typically the latter is followed by flocculation with polymer aids. Figure 4.4.3 shows a labyrinthine-type flocculator in a textile dyeing wastewater treatment plant. Such flocculators avoid the need for low speed stirrers and may be favored if there is a desire to reduce the number of mechanical elements in the IWTP. The hydroxide precipitate formed from the coagulant's reaction with

Fig. 4.4.3. Labyrinthine-type flocculator (LHS) at a IWTP for a textile dyehouse. Dosing of the flocculant from a pipe just before the start of the first baffle of the flocculator can be seen on the LHS of the picture.

alkalinity agglomerated into larger more settleable particles in the flocculator, and liquid-solids separation would then take place in the clarifier (Fig. 4.4.4). The latter is of the rectangular configuration as opposed to the circular configuration. The rectangular configuration usually lends itself better to a more space-saving arrangement of vessels at space constrained sites.

The process of coagulation can also assist in removing the dyes dissolved in the textile wastewater. The use of coagulants to remove colors in industrial wastewater is a frequent occurrence. For such applications, iron salts may perform better than alum. As in the use of coagulants to assist the removal of O&G and fine particulates, coagulant assisted color removal also has the problem of disposing large quantities of sludge.

It should, however, be noted that not all wastewaters are coagulated and flocculated before clarification. Pig farms generate very strong wastewaters in terms of suspended material and dissolved components. Early removal of the suspended fraction serves to improve the performance of downstream biological unit processes. Such removal may be achieved with clarifiers. Notwithstanding this, and in the case of piggery wastewater and other easily biodegradable wastewaters, the use of clarifiers has not always been successful. This is because septic conditions can easily develop if hydraulic retention times become too long resulting in the

Fig. 4.4.4. Rectangular clarifiers following coagulation/flocculation at a textile dyehouse. Rectangular vessels may be easier to arrange in a more space-saving manner compared to circular vessels.

Fig. 4.4.5. Drum-type mechanical fine screen at a pig farm. The screenings below the mechanical screen are not biologically stable and have to be further treated. This may involve composting or liming.

release of gases generated. These then interfere with the settling process and possibly resulting in rising sludge. Such gases are often odorous and this would certainly be so in the case of piggery wastewater. To avoid this, the primary clarifier can be replaced with the fine screen although the latter does not quite match the performance of a clarifier which is operating well. Figure 4.4.5 shows a mechanical fine screen which has been applied to piggery wastewater in place of the primary clarifier. Screens can be an effective alternative to clarifiers where space at a site is constrained. Such fine screens may be mechanical as shown or non-mechanical versions like the curved self-cleaning screens. The non-mechanical versions which do require more attention from operators to avoid clogging, and hence overflowing, have been used successfully at locations where manpower is readily available and inexpensive.

Apart from the obvious reduction in solids load on the bioprocess by clarifiers and fine screens, the latter could also aid the performance of downstream processes like pH adjustment. An example of this is pineapple canning wastewater. During preparation of the fruits prior to canning, trimming and washing result in bits of the fruit being carried away in the wastewater. Since the fruit is acidic, the bits of fruit making up the particulates are acidic and would have consumed large quantities of alkali if pH adjustment was attempted in their presence. These particulates can be easily removed with fine screens and their removal improves performance of the pH adjustment stage particularly in terms of alkali consumption.

## 4.5. pH Adjustment

Unlike domestic or municipal sewage where the pH range is typically $6.0 \sim 7.5$, industrial wastewaters have pHs which vary over a much broader range — from very acidic to very alkaline. It should also be noted that a factory may generate a number of wastewater streams and among these can be those which are acidic while the rest may be alkaline.

Consequently it can be useful in terms of reducing chemical consumption for pH adjustment by providing sufficient equalization prior to pH correction so that the various wastewater streams may achieve a degree of pH adjustment through their own interaction. This becomes particularly important if the acidic and alkaline streams are not generated at the same time. Holding and blending becomes a necessary activity then. The difficulties with pH need not always occur because it is an inherent characteristic of the wastewater. It should be noted that pH may be manipulated if chemical cracking of oily emulsions and coagulation had been necessary (as discussed in Sec. 4.4) and may subsequently need to be adjusted again prior to biological treatment. Automatic pH correction can be an unexpectedly difficult activity to perform satisfactorily. This is, in part, because of the difficulty associated with mixing a small quantity of reagent uniformly with a large volume of wastewater. This is made even more difficult if wastewater characteristics, such as its flowrate, changes rapidly. The value of adequate equalization or blending prior to pH adjustment cannot be overstated.

Because of the relatively small size of many IWTPs, the preferred chemical for pH adjustment of acidic wastewaters is usually sodium hydroxide instead of lime. A solution of sodium hydroxide would be prepared prior to its injection into the pH correction tank. At IWTPs where the chemical consumption is sufficiently large to justify the additional handling facilities required, lime made up in the form of a slurry can be used. The handling of lime powder (its typical form when delivered to the IWTP) has safety requirements which operators at small IWTPs may not be equipped to cope with. Lime is usually chosen because it is cheaper than sodium hydroxide. When lime is used, it is necessary to appreciate that it is slower compared to sodium hydroxide. This means that the reaction tank has to be increased in size to allow for the longer hydraulic retention times needed. Typically a minimum of 20 min HRT is allowed for. The reaction tank's contents are mixed either with a mechanical stirrer or with air.

Where alkaline wastewaters need to be pH corrected, the chemical frequently used is sulphuric acid. The reason for the choice is again usually cost. If the downstream processes include an anaerobic process and relatively large quantities of acid are required, then sulphuric acid may be substituted with hydrochloric acid. This is because the sulphates can be reduced in the anaerobic process resulting in odorous and corrosive hydrogen sulphide being released with the gaseous emissions from the anaerobic reactor.

Figure 4.5.1 shows a pH correction station at a pharmaceutical factory. While not obvious from the figure, this is a two-chambered pH correction station designed to deal with the high pH of the incoming wastewater. The latter had been 9.5 and higher on several occasions. It should be noted that the relationship between pH and the reagent flow required to bring about change is highly non-linear. This arises because of the logarithmic nature of the pH scale. A change of one pH unit means a ten-fold change in acidity or alkalinity. The tank's two chambers were arranged in series with the first smaller than the second. Two stage pH adjustment was practiced with pH in the first chamber being adjusted over a

Fig. 4.5.1. An excavated pH correction station (LHS) with landscaped surroundings. IWTPs are often constructed very close to a factory and in the design and location of major structural components, like the vessels, some consideration may have to be given to the aesthetics of the IWTP in relation to its surroundings.

relatively broad band with a larger dosing pump while pH adjustment in the second chamber was over a much narrower band with a smaller dosing pump. This approach was necessary to avoid the swings in pH values if a single adjustment chamber with a large dosing pump had been used.

The vessels used for the various unit processes in an IWTP may be built as fully excavated, partially excavated, or at ground level. Often the decision as to which level a vessel should be placed depends on the desired hydraulic grade line so that flow through an IWTP can be maintained, insofar as is practicable, without the aid of further pumping after the lift at the start of the plant. While the preceding may indeed be a frequent criterion for deciding the level, it need not always be so. The pH correction station shown in Fig. 4.5.1 had excavated vessels. The primary reason for doing so in this instance was to maintain a level of aesthetics acceptable to the owner.

Many IWTPs include a bioprocess in their treatment train. Bioprocesses are sensitive to pH conditions and operate well within the fairly narrow range of $6.5 \sim 7.5$. While pH conditions outside of this range need not necessarily result in a toxic condition, the bioprocess may nevertheless be inhibited or certain species of micro-organisms may be favored over the more desirable species. One of possible consequences of the latter phenomenon is bulking sludge. Failure to adequately control pH has been noted to be the cause of a surprisingly large number of IWTPs failing to produce wastewater of the required quality from both the bioprocesses and physico-chemical processes.

## 4.6. Removal of Inhibitory Substances

Since many IWTPs include a bioprocess in its treatment train, the presence of potentially inhibitory substances is of concern as the performance of the plant and hence the ability to meet the discharge limits may be adversely affected. These potentially inhibitory substances can include organics, metals, and substances such as O&G, ammonia, and fluoride. Successfully matching a bioprocess to a wastewater requires the bulk of the organics present in the wastewater to be biodegradable. If the wastewater is inhibitory, then this should usually be caused by substances other than the bulk of the organics. To protect the bioprocess, these potentially inhibitory substances would have to be removed from the liquid stream before it gets to the former.

A combination of pH adjustment, precipitation, and coagulation is frequently used in IWTPs to remove these substances and this is especially so for the removal of metals. For example, rubber glove factories prepare a latex formulation which included zinc in preparation for injection into molds on the production line. This results in wastewater which contained latex, organic acids, and zinc. While the bioprocess can be expected to remove organic acids, the latex would interfere with the treatment by imposing a large, difficult-to-degrade organic load on the bioprocess while the zinc can be at concentrations which are potentially inhibitory. A combination of pH adjustment to achieve zinc precipitation, as well as coagulation to assist removal of the zinc precipitate and latex, and dissolved air flotation to assist liquid-solids separation has been successfully used. Figure 4.6.1 shows such a plant at a rubber gloves factory. While the coagulation did reduce overall organic content by removing much of the latex, residual organic content was still high in this instance because the organic acids concentration had been high in the raw wastewater and this dissolved organic component was not significantly affected by the coagulation process.

Removal of inhibitory substances by precipitation, such as the zinc cited in the preceding example, results in a sludge which can be classified as a potentially toxic metal sludge. Disposal of such sludges after dewatering should be at controlled landfill sites. This can be a significant cost component in the overall cost structure of a IWTP's operation.

Fig. 4.6.1. A physico-chemical plant for zinc and latex removal. Precipitation (for zinc removal) and coagulation-flocculation (for zinc and latex removal) preceded the DAF (elevated rectangular tank on the RHS). The DAF separated solids generated from the liquid stream. The resulting liquid stream was further treated with bioprocesses.

The use of activated carbon adsorption to remove potentially inhibitory organics prior to the bioprocess is rare. The reason for this is the presence of large quantities of non-inhibitory organics in the wastewater competing for the activated carbon's adsorption sites. This would lead to inefficient, and costly, use of the adsorbent. Activated carbon, where it is used, is usually after the bioprocess where the former serves as the polishing step for removal of small quantities of persistent organics. The latter may include color caused by textile dyes.

## 4.7. Nutrients Supplementation

In sewage treatment, nutrients removal is becoming an increasingly common requirement but in industrial wastewater treatment this need not necessarily be so. While it is true there can be wastewaters (eg. slaughterhouse wastewater which contains blood) which may also require nutrients removal (eg. nitrification and denitrification of slaughterhouse wastewater to remove nitrogen), many may, in contrast, require nutrient supplementation to support a healthy bioprocess. This is because industrial wastewaters may have organic contents which are too high in relation to their nitrogen and phosphorus content (ie. the BOD:N:P ratio of 100:5:1 is not met because the BOD component exceeded 100). The common chemicals used to supplement nitrogen and phosphorus are urea and phosphoric acid. These are typically added after the processes discussed before this section, if any, have been carried out so as to avoid losses and interference to the former. The nutrients can be dosed into the pretreated wastewater as it is conveyed to the bioprocess or dosed directly into the bioreactor. The configuration shown in Fig. 4.7.1 is an example of the former and arose because of space constraints at the site.

In small IWTPs, ammonium dihydrogen phosphate may be used to avoid having to handle phosphoric acid. Due to its cost compared to urea and phosphoric acid, this compound is usually used only if the nutrients supplementation requirement is relatively small. On similar arguments of cost and chemicals handling, and where it is available, domestic sewage has been blended with industrial wastewater to create a more balanced wastewater in terms of BOD:N:P. This option can be explored if a factory has a dormitory for its workers.

While only nitrogen and phosphorous supplementation would often be adequate, there are instances where these macro-nutrients, on their own, are insufficient. This situation, however, rarely occurs when dealing with agricultural or agri-industrial wastewaters but can occur where a factory's processing or manufacturing activities do not involve natural materials or the factory's inputs are of highly processed materials leading to the absence of certain micro-nutrients.

Fig. 4.7.1. An in-line nutrients supplementation arrangement where the dosing pump injected a nutrients solution into the pipe conveying pretreated palm oil refinery wastewater to the bioreactor.

These micro-nutrients include Mg, K, Ca, Fe, Mn, Cu, and Co. An example would be the chemical industry wastewaters which require magnesium to be supplemented, at least on occasion, to support formation of sludge granules or flocs with sufficient density. This allows for better operation of the anaerobic processes in the treatment train. A less obvious example can be soft drinks wastewater. Although a food industry wastewater, this wastewater's composition can be relatively well defined (according to the drinks' formulations). The use of nitrogen and phosphorus supplements alone allows the bioprocess to work but operation can be difficult in terms of process stability and selection of a microbial population which allows for better flocs formation and therefore easier clarification and excess sludge management. The application of a micro-nutrient, potassium, was noted to have eased the problem.

Apart from the macro- and micro-nutrients, biocatalyst additives are now commercially available. These additives usually contain mixtures of enzymes, dried bacterial solids, and possibly by-products of bacterial fermentation. Such additives are often recommended to owners of IWTPs to improve the operation of the bioprocesses therein and hence improve effluent quality. While such additives may indeed work, continuous or regular application is usually required to maintain the performance. The consequent substantial operating costs incurred may well mean alternative, more "permanent", and economical solutions for the bioprocess difficulties should be explored. Such additives do, however, have a role to play as a "quick fix" while process difficulties are investigated and remedied.

# CHAPTER 5

# THE INDUSTRIAL WASTEWATER TREATMENT PLANT — BIOLOGICAL TREATMENT

## 5.1. Microbes and Biological Treatment

Like a sewage treatment plant, secondary treatment processes follow the preliminary and primary processes in an industrial wastewater treatment plant. Typically the latter processes would not have been able to produce a treated effluent which can meet the discharge limits since they primarily target the settleable or floatable pollutants. Colloidal and soluble pollutants can be expected to penetrate the preliminary and primary processes and often these pollutants are organic in nature and biodegradable. Hence primary treated wastewaters would need further treatment and such secondary treatment is often biological in nature. However, unlike STPs, these bioprocesses need not always be aerobic in nature and need not provide almost complete stabilization in a single stage. In industrial wastewater treatment, the bioprocesses can be used to provide partial and subsequently almost complete stabilization of the biodegradable substances. There are many examples of anaerobic processes providing partial stabilization before further treatment with aerobic processes. There are also many examples with the latter as the only biological component in the treatment train. The reason for staged treatment is the relatively high organic strength of many industrial wastewaters. These high organic levels can make aerobic treatment on its own difficult to achieve technically and/or economically. Consequently anaerobic processes may be used to reduce wastewater organic strength prior to aerobic treatment. This is in contrast to STPs where the anaerobic process is typically used to digest the organic sludge generated by the primary and secondary processes, but is rarely used to treat the liquid stream.

Both the aerobic and anaerobic processes depend on microorganisms to provide the functional basis for the treatment processes which include carbon oxidation, nitrification and denitrification, acidogenesis, and methanogenesis. Bacteria are the micro-organisms of principal interest and the bulk of these would be the heterotrophs — organisms which use organic carbon for cell synthesis.

Notwithstanding this, there are autotrophs, organisms which use inorganic carbon for cell synthesis (eg. carbon dioxide), that are important to wastewater treatment. An example is the nitrifiers which convert ammonia to nitrate in the nitrification process. Although treatment processes are generally identified as aerobic and anaerobic, the bacteria in the "aerobic" processes are in fact largely facultative. Such bacteria are not obligate aerobes but are able to function under both anaerobic and aerobic conditions. Aside from their importance in the removal of carbonaceous pollutants, facultative bacteria are also important in denitrification where combined oxygen in nitrites and nitrates are removed releasing nitrogen gas. Obligate aerobes would have required the presence of molecular oxygen to thrive. The anaerobic processes in contrast depend on many obligate anaerobes and these can only thrive in the absence of molecular oxygen. The presence of facultative microorganisms in anaerobic systems and their ability to utilize combined oxygen, as in sulphates for example, gives rise to problems such as odor, corrosion, and toxicity (sulphates being reduced to hydrogen sulphide).

These aerobic, facultative, and anaerobic bacteria have several distinct shapes or morphology. Many bacteria species common and important to wastewater treatment belong to the cocci or spherical shaped bacteria and the bacilli or rod-shaped bacteria. A biological reactor would have a community of bacteria made up of a mixed culture. It is usually desirable, in wastewater treatment, to have such mixed cultures so that a wide range of pollutants can be handled. A mixed culture is also likely to be more robust when challenged with changing wastewater characteristics. Bacteria cells in the population secrete a slime layer which is made up of various organic polymers. It is believed this slime layer is the key to microbial flocculation allowing the cells to agglomerate, forming more settleable floc particles and hence resulting in more effective gravity liquid-solids separation in the clarifiers. Slime layer formation is more significant for "older" cultures (ie. stationary phase and beyond, or processes with longer cell residence times or CRTs). Cultures which are in the growth phase typically have less extensive slime layers and flocculation is therefore weaker. A possible consequence of this is the formation of "pin-head" flocs (ie. very small flocs) or dispersed growth and turbid effluent following clarification.

Apart from the bacteria, the microbial population can also be expected to include unicellular organisms such as protozoa and even higher plants and animals such as rotifers and crustacea. All these are aerobic in nature and their presence can only be expected in adequately oxygenated systems. Aside from serving as indicators of a healthy biological process, such organisms serve as grazers of the bacteria population and hence help reduce excess sludge yields. A microbial population may also include fungi. The latter may also be unicellular or multicellular and like bacteria it can play an important role in wastewater treatment although this can be a largely negative role. Fungi tend to compete better than bacteria at lower pH and nutrient deficiency conditions. A shift of species dominance from bacteria to fungi (which can be filamentous) can result in bulking sludge. Fungi can also proliferate when treating wastewaters with relatively simple organic pollutants because of an apparent nutrients deficiency condition.

The size of the culture in a reactor is an important design consideration as it is determined by the amount of pollutants which has to be converted when treating the wastewater. This is typically represented by the amount of organic suspended material in the mixed liquor of a reactor or the mixed liquor volatile suspended solids (MLVSS). It is assumed that the organic suspended material is largely made up of microbes. The loading on a reactor is typically defined as the mass of substrate applied on unit mass of MLVSS over a defined period of time (eg. $0.3 \, \text{kg BOD}_5 \, \text{kg}^{-1} \, \text{MLVSS} \, \text{d}^{-1}$). Upon considering aerobic processes, such as the activated sludge process, this loading is commonly referred to as the F:M ratio (Food to Microbial mass ratio). The MLVSS is therefore an important parameter to monitor when operating a bioprocess to ensure an adequate population of microorganisms is retained in the reactor to perform the necessary functions and the process does not become overloaded. MLVSS can become depleted because of microbial cell washout, inappropriate desludging protocols, and inhibition. For operators who become familiar with their plants and wastewater, an alternative to the MLVSS is the MLSS. The latter requires less effort to determine and although it includes inorganic material which may not be of microbial origin, coefficients can be determined and applied on it to obtain reasonable estimates of MLVSS.

The microbial culture considers the bulk of the pollutants in a wastewater as substrates. These substrates would be metabolized and the metabolic reactions involved are very largely enzymes driven. As the bacteria metabolizes the organic substrates, they reproduce by binary cell division and the time required by various bacteria to prepare for such division may range from minutes to hours (and hence the cell doubling time). Cell reproduction, in addition to the carbon sources which are the carbonaceous pollutants, also requires the presence of nitrogen and phosphorous, and hence the need to supplement these macro-nutrients should there be a deficiency in the wastewater. The microbial mass increase in a culture as a consequence of such cell division is the microbial yield. This adds to the MLVSS and has to be removed either continuously or at regular intervals by desludging the excess sludge. Typically this would be from the secondary clarifiers for continuous flow aerobic systems and from the reactors for the batch and anaerobic systems. Enzyme activity in the metabolic reactions is dependant on environmental factors such as temperature and the presence of metal activators. Many bacteria thrive best at temperatures of 25–40°C and these are the mesophiles while those which are better suited to higher temperatures, 55–65°C, are the thermophiles. Most enzymes become progressively denatured at temperatures above 65°C. Damage to the enzymes causes the metabolic reactions to slow and then stop, and consequently process failure. A third group of microbes, the cryophiles, which thrive at temperatures below 20°C (typically at 12–18°C) is rarely utilized in wastewater treatment. Most wastewater treatment systems depend on the mesophiles as large numbers of the wastewaters to be treated are at ambient temperature and in many parts of Asia, the latter would be in the range of 22–36°C. Enzyme activity is also affected by the presence or absence of metal activators. The necessity for these metal activators accounts for the necessity to supplement micro-nutrients in some industrial wastewaters. Fortunately most wastewaters do not require such supplements because the micro-nutrients would have been inherently present in adequate quantities — often times because these are contaminants in the raw materials used in the manufacturing process. The absence of adequate quantities of macro- and micro-nutrients can result in the development of filamentous microbial cultures and the phenomenon called bulking sludge. The latter is undesirable as it may affect liquid-solids separation in the clarifiers and subsequently in the sludge dewatering process as the consequent sludge structure allows it to retain more water. Macro- and micro-nutrients deficiency is not an issue which is of concern at STPs. Excess quantities of macro-nutrients is, however, becoming an issue at STPs and at some IWTPS, and this calls for nutrients removal (ie. N and P removal). Excess micro-nutrients or metal salts are usually not issues of concern at STPs. However, excess metals salts can be an issue at IWTPs. This is particularly so with sodium chloride (NaCl), a salt which can be present at relatively high concentrations if a batch of industrial wastewater included the spent regenerant stream from ion exchangers. Substances are transferred in and out of a bacteria cell by osmosis. If the concentration of electrolytes outside of the cell is greater than that inside, water migrates out of the cell in an attempt to restore equilibrium. However, should the electrolyte (eg. NaCl) concentration outside the cell is so high and this migration continued, the bacteria cell would undergo plasmolysis and in effect dehydrate leading to the cell's destruction eventually. As a gross measure of the salts, the parameter Total Dissolved Solids (TDS) may be used. For anaerobic systems, a value as low as $1500\,\text{mg}\,\text{L}^{-1}$ may already cause noticeable adverse effects on the bioprocess. For aerobic systems, TDS would typically be kept below $3000\,\text{mg}\,\text{L}^{-1}$ and while it may be possible to operate adequately acclimated systems receiving wastewaters with much higher TDS concentrations, values above $4500\,\text{mg}\,\text{L}^{-1}$ should be approached with caution. At TDS values above $10\,000\,\text{mg}\,\text{L}^{-1}$, aside from potential difficulties with the bioprocess, there may also be difficulties with gravity liquid-solids separation.

## 5.2. Measures of Organic Strength and Oxygen Demand

One of the primary determinants in the sizing of a bioprocess is, of course, the organic content which has to be removed from the wastewater. This organic content can be indicated by any one or combination of the following:

(i) Biochemical oxygen demand (BOD);
(ii) Chemical oxygen demand (COD);
(iii) Total organic carbon (TOC).

The BOD is the most widely used and is a measure of the oxygen required to stabilize a sample's biodegradable organic content. This would typically be the result of a 5 day test conducted at 20°C and reported as $BOD_5$. The BOD test is, however, not without defects. In industrial wastewater treatment, the facility designer and operator should be aware that $BOD_5$ results can be very different between tests using acclimated and non-acclimated seeds. The former would be better able metabolize the organic substances and hence yield a higher $BOD_5$. When the BOD is used as an indicator of the biodegradable organic content for sizing aeration equipment, the ultimate BOD ($BOD_{ult}$) must be used. This is so because the $BOD_5$ only measures the amount of oxygen consumed by micro-organisms during the first 5 days of biodegradation and this is not the total amount of oxygen required by microorganisms to oxidize the biodegradable organics to carbon dioxide and water. $BOD_{ult}$ is typically approximately 1.5 times $BOD_5$. With the recognition that some industrial wastewaters may have significant quantities of Ammonia-N present, it should be noted that the BOD test also does not allow sufficient time for the Ammonia-N oxidation process. Again for purposes of sizing the aeration equipment, and where nitrification is a design requirement (because of the need to maintain relatively long sludge ages), the Nitrogenous BOD has to be included in the calculations and this is approximately 4.6 times the total of the Ammonia-N and the Organic-N content. Nitrogenous BOD or the $BOD_N$ can contribute to a significant portion of the total oxygen requirement. Unless otherwise stated, it should be assumed the $BOD_5$ values provided do not include the $BOD_N$ and allowance should then be made for the latter in the overall process design.

The COD is also a test for determining oxygen demand but does not depend on the ability of micro-organisms to degrade the organic substances in the wastewater. The latter may be recalcitrant or even toxic to micro-organisms. The use of a strong chemical oxidizing agent, potassium dichromate, makes the COD test much quicker than the $BOD_5$ test — hours compared to days in the latter case. The dichromate COD value is usually higher than the $BOD_5$ value of a given sample. The COD:$BOD_5$ ratio can be used as a first indicator of the biodegradability of the wastewaster's organic substances and hence the potential suitability of a bioprocess for inclusion in an industrial wastewater treatment plant. Typically COD:$BOD_5$ values greater than 3.0 would suggest that application of a bioprocess should be approached with some caution. After a plant has commenced operation, operators may initially conduct both the COD and $BOD_5$ tests regularly and develop a relationship between the two sets of results. Subsequently the COD, being the faster test, can be used in routine monitoring of wastewater quality and plant performance. On occasions a COD value which is numerically more similar to the $BOD_5$ value of a sample may be encountered. This may refer to the permanganate COD value (or PV) and in some cases it is used as the equivalent of a sample's $BOD_5$ value. The oxidizing agent in this instance is not potassium dichromate but potassium permanganate which is a less aggressive oxidizing agent.

The TOC is, compared to the $BOD_5$ and COD, less frequently encountered in industrial wastewater treatment. This is because many factories do not report it since regulatory agencies typically call for reports on $BOD_5$ and COD to determine compliance with the discharge limits. The TOC is determined with a TOC analyzer which converts organic carbon to carbon dioxide using either a strong chemical oxidizing agent or by combustion in the presence of catalysts. The carbon dioxide generated is then measured with an infra-red detector. Should the instrument be available, it can be used like the COD to develop a relationship with $BOD_5$ (or COD). Since the TOC is even faster than the COD test (minutes compared to hours in the COD), it can be used to provide early warning of changes in wastewater quality allowing plant operators more time to institute emergency response actions. The TOC, unlike the $BOD_5$ and COD, is a measure of the organic content in the wastewater and not a measure of the oxygen demand caused by this organic content. Unfortunately, somewhat like the COD, it does not provide an indication of the biodegradability of this organic content.

It is important to note that the bioprocesses in a treatment plant have the primary objective of removing the organic substances from a wastewater. However, these organic substances need not all be similarly biodegradable. There can be substances present which are more difficult to remove biologically, and there may even be some which are resistant to biodegradation unless appropriate conditions are present and hence are persistent. Examples of difficult-to-degrade substances include oil and greases, textile dyes, phenols, tannic acid, lignins, and cellulose. A bioprocess can therefore be expected to remove only the biodegradable fraction of the organic substances present, perhaps the persistent substances, and certainly not the recalcitrant substances. So for the less or non-biodegradable fraction, unless provisions have been made for their removal, these would remain in the treated wastewater and exit the plant as residual organics. Given this, an IWTP dependant on a bioprocess for organics removal may be able to satisfy the discharge limit for $BOD_5$ but may not necessarily do so for COD.

## 5.3. Reactor Configurations

Because wastewater is typically thought of as being discharged continuously over the working day, it is generally assumed that it would be more convenient to design continuous-flow reactors which can handle such continuous flows. Many, but not all, bioreactors in IWTPs are indeed of the continuous-flow type. There is, however, a significant number which are not continuous-flow reactors but are batch (or cyclic) reactors. The choice of the type of reactors to use depends on the flow pattern of the wastewater and the bioprocess selected in response to the latter's characteristics. The wastewater flow pattern is a particularly important consideration since a wastewater is rarely discharged at the same flow rate throughout the working day (see Chapter 2).

There are three types of reactors commonly used in industrial wastewater treatment. Two of these are continuous flow types — the continuous-flow stirred-tank reactor (CFSTR) and the plug flow reactor, while the third is the batch reactor. The CFSTR assumes perfect mixing throughout the reactor. This confers two important features to the reactor — as pollutants flow into the reactor, their concentrations are instantaneously diluted to the concentrations in the mixed liquor, and secondly the pollutant concentrations in the reactor effluent are the same as the concentrations in the mixed liquor. A consequence of the latter feature is the relative absence of a pollutant concentration profile in the longitudinal direction (ie. overall flow direction) within the reactor. This means that the provision of supporting systems, such as the aeration system, would be on the basis of homogenous conditions through the reactor and is hence relatively simple. However, the difficulty with this configuration is that most IWTPs receive an inflow which varies in terms of flow pattern and composition, and these variations can occur relatively quickly. Since the treated effluent must consistently meet the discharge limits, the process must then be designed for the maximum loading rate. This is a consequence of assuming steady-state conditions for purposes of process design in most cases and the necessity then to have a relatively slow rate of change. The extension to this then is the necessity to have relatively long hydraulic retention times and hence large reactor volumes. CFSTR vessel configurations are typically circular or square and process variants having the complete mix regime includes the complete-mix aeration and extended aeration (do note the extended aeration mode as in oxidation ditches is plug flow instead).

To improve the situation described above in terms of reactor volumes, a cascade of CFSTRs in series instead of a single large one could be attempted. Theoretically, it may be shown that the larger the number of these smaller CFSTRs, the smaller the total tank volume would be compared to the case of the single large CFSTR. The reason for this phenomenon is the higher average mixed liquor substrate concentration in each of the smaller CFSTRs as opposed to that in the single large CFSTR producing a final effluent of similar quality. Because of these higher concentrations (compared to the effluent concentration), the average reaction rate would also be higher in the cascade and hence the reaction time required can be shorter. The ideal cascade is one having an infinite number of smaller CFSTRs but in practice this cannot be realizable and a more typical number of these smaller tanks or stages in industrial wastewater treatment would be three and very rarely more than five.

The cascade of CFSTRs presents a situation where mixing is ideal within the individual reactor but there is no mixing as the wastewater moves from one reactor to the next. This is somewhat similar to a situation where mixing is ideal in the lateral plane but absent in the longitudinal plane — ie. plug flow. In reality, ideal plug flow cannot be achieved and neither can the ideal CFSTR. Reactors would have dead volumes, short circuiting, and dispersion (where material is transported from regions of high concentration to regions of low concentrations by turbulence in the reactor). Notwithstanding this, plug flow reactors should be smaller than equivalent CFSTRs. Although the plug flow reactor may have the advantage of a smaller size, there is the difference in the mixed liquor characteristics in the longitudinal direction to contend with. An example is the pollutant concentration and hence development of an oxygen demand gradient in the longitudinal plane. Accommodating the latter requires the installation of an aeration system which provides more oxygen at the head of the reactor than at the discharge end. Plug flow reactors are typically long, narrow vessels. Many activated sludge process variants such as contact stabilization, tapered aeration, step-feed, conventional, and oxidation ditch processes have plug flow regimes.

In biological wastewater treatment systems, the processes held in CFSTR and plug flow reactors are expected to be operating at steady-state. Plug flow reactors operating at steady state have mass balances which are similar to those for batch reactors although the latter are always non-steady state. It can be shown that the batch reactor with instantaneous fill would have the same volume as an equivalent ideal plug flow reactor. However, an instantaneous fill in wastewater treatment is not practicable. Consequently batch reactors with finite fill times have volumes larger than the ideal plug flow reactor although they would still be substantially smaller than the CFSTR. Since the plug flow and batch reactors have similar mass balances, it can be expected that the batch reactors would also have mixed liquor characteristics similar to that found in plug flow reactors. The difference in this instance is that the gradients occur over a temporal frame instead of the spatial frame of the plug flow reactor. The batch reactor is therefore a complete mix reactor at any given point in time. Batch reactors, like those in the sequencing batch systems, are much less sensitive to reactor configuration than the CFSTRs and plug flow reactors. While the batch reactor vessels in industrial wastewater treatment are often square or circular, there have been occasions when they are more unusually shaped. Figure 5.3.1 shows a "trapezoidal" looking vessel for a sequencing batch reactor (SBR) being constructed for a food factory. This configuration had resulted because of the configuration of the space available for construction.

Unfortunately the differences in actual reactor volumes are usually not as large as suggested by the theoretical considerations discussed above. This is because the

Fig. 5.3.1. The "trapezoidal" tank under construction in the foreground was to be a SBR vessel. The SBR was selected because it is least sensitive to vessel configuration. Available space at the site did not permit construction of vessels with "regular" shapes and still achieve the required hydraulic capacity.

ideal CFSTR, plug flow reactor, and batch reactor are not achievable in practice — particularly so for the CFSTR and plug flow reactor. In reality, the conditions within either one of these two reactor configurations are somewhere between the two "extremes" and this is referred to as arbitrary flow. Consequently there is a tendency to be conservative when applying the reaction rates and hence in effect over-design. The "over-design" may also come about because of the necessity to cope with wastewater characteristics which are far from stable. All three types of reactors or, more likely, nominal versions of these are used in industrial wastewater treatment. The decision to use one type of reactor or the other depends not only on its possible size but also on each reactor configuration's ability to cope with wastewater characteristics, the equipment available which can cope with operating conditions (eg. DO profile) within each type of reactor configuration, and the anticipated ability of the plant operators to cope with the characteristics of systems designed around each of these reactor types.

## 5.4. Suspended and Attached Growth

The microbial community in the reactors described above can be in suspended or attached growth form (ie. biofilms). Reactors often have one or the other but on occasion can have a combination of suspended growth and biofilms. In suspended growth systems, the microbial mass is suspended in the reactor's mixed liquor as flocs. These flocs are kept in suspension with agitation by mechanical mixers or gas injection. The latter would typically be air in aerobic systems and biogas in anaerobic systems. The agitation facilitates intimate interaction between substrates and biomass. In biofilm systems, microbes attach themselves, in a thin layer, onto a support medium. The latter may be in the form of a fixed bed or moving bed. Fixed or stationary beds are typically of moulded plastic shapes or gravel while moving beds can comprise granular activated carbon or sand grains. These beds of support medium may be submerged in the mixed liquor throughout the reactor's operation or alternatively exposed to air and wastewater.

The bulk of the aerobic systems used in IWTPs are suspended growth systems. Examples of these include the aerated lagoon, conventional activated sludge process, oxidation ditch, and sequencing batch reactor. Occasionally such aerobic systems may incorporate biofilms with suspended growth. These may involve fixed bed biofilms as in the activated sludge system shown in Fig. 3.4.1. Such reactors then provide both suspended and stationary contact between the substrates and microbes. The biofilm can be formed on blocks of support medium assembled out of a plastic netting material as shown in Fig. 5.4.1. Biofilms may also be formed on freely suspended support medium made up of small particles or plastic shapes.

Fig. 5.4.1. Plastic netting-like material which can be assembled into blocks of support medium for biofilm formation.

An example of a material which has been used as the freely suspended support medium is powdered activated carbon.

Somewhat less frequently encountered in IWTPs are the aerobic biofilm systems known as the trickling filter and rotating biological contactor (RBC). RBCs, although encountered in sewage treatment, are rarely encountered in industrial wastewater treatment in Asia. Trickling filters which are non-submerged fixed support medium systems are, comparatively, more frequently encountered but are still rare compared to the suspended growth aerobic systems. Trickling filters have been used to provide preliminary treatment of strongly organic wastewaters (Fig. 5.4.2). The trickling filter then serves as a roughing filter (ie. for organic strength reduction). Given the high organic loads, in order to provide sufficient oxygen to the biofilm in the trickling filter, forced aeration may be used. The air is forced, countercurrent to the downflowing wastewater, through the support medium. The roughing filter's effluent would be further treated, usually with an aerobic suspended growth system.

In anaerobic systems, the biomass is rarely applied in the form of flocs suspended throughout the volume of mixed liquor. Although rare the latter configuration does occur in the treatment of certain types of wastewaters. An example is palm oil mill effluent (POME) where the anaerobic reactors can be of the complete mix type. Instead the more frequently encountered configuration has the

Fig. 5.4.2. Roughing trickling filter with forced ventilation at a coffee extraction plant.

biomass flocculated or granulated and retained in the reactor in the form of a sludge blanket. Examples of systems using such an approach include the upflow anaerobic sludge blanket (UASB) and anaerobic sequencing batch reactor. Where biofilms are applied, this is usually by way of a fixed support medium. The latter may be rigid moulded plastic shapes or flexible plastic fibres (Fig. 5.4.3) in the form of "ropes" or rings strung on steel wires. This configuration gives rise

Fig. 5.4.3. "Fibrous" support medium mounted within an anaerobic filter (view from the top).

to the anaerobic filters and where the sludge blanket has been combined with the stationary biofilm, the hybrid anaerobic reactor.

## 5.5. Anaerobic Processes

The anaerobic process comprises a series of interdependent phases. Initially complex organic compounds such as lipids, proteins, and carbohydrates, if present, are hydrolyzed to simpler organics. The latter are then fermented to volatile fatty acids (VFAs) by acidogens. The most common of these fatty acids is ethanoic acid. However, propanoic, butanoic, and pentanoic acids may also be present in varying quantities depending on the stability of the process. Given the production of acids by the process, the system has to be adequately buffered to avoid pH declines which may adversely impact on the process's further progress. The acidogens include both facultative and obligate anaerobic bacteria. Up till this point in the process, the total amount of organic material persent in the wastewater would not have changed significantly although the type and complexity of organic compounds could have changed substantially. The gaseous by-product of the acidogenic reactions is carbon dioxide. Subsequent to the acidogenic phase is the methanogenic phase. The methanogens are obligate anaerobes and they convert the fatty acids from acidogenesis to methane and carbon dioxide. This results in substantial decrease in the organic content of the wastewater. The methane generated offers an avenue for energy recovery.

The anaerobic process is a complex process and there is substantial opportunity for it to become unstable and eventually fail. Among the important environmental conditions which should be present is the absence of molecular oxygen. This is particularly so for high rate processes. Such anaerobic systems should be designed with reactors which have positive pressure within the vessels so that air is excluded. An indication of impending anaerobic process failure is dropping pH. The methanogens are sensitive to pH and methanogenesis would stop if pH drops below 6.2. Bearing in mind acidogenesis precedes methanogenesis, pH control is an important consideration in the operation of anaerobic systems. The microbial consortium in an anaerobic reactor also needs an appropriate balance of macro- and micro-nutrients to ensure microbial growth can occur. However, anaerobes have relatively slow growth rate and in sludge digestion this is a desirable characteristic as it meant low solids production. The methanogen cell yield is lower than the acidogen's. The low biomass yield does mean the nutrients requirement of an anaerobic process is lower than that of the aerobic process. In terms of BOD:N:P, the aerobic process would have required 100:5:1 while the anaerobic process only require 100:3.5:0.5. Notwithstanding this lower requirement, nutrients supplementation may still be necessary since many industrial wastewaters are nutrients deficient even for anaerobic processes.

In wastewater treatment, the low microbial cell yields can be a hindrance since slow growth rates mean the necessity for long hydraulic detention times unless hydraulic retention times (HRT) can be effectively separated from cell residence times. It also means anaerobic processes may not be as effective as aerobic processes when faced with high hydraulic loads which are accompanied by relatively low organic loads. Since industrial wastewaters can have organic strengths very substantially higher than sewage, anaerobic processes can be used to treat the liquid stream and this would take place ahead of the aerobic processes. This is in contrast to sewage treatment where the anaerobic process is usually used to digest the primary and secondary sludges at the end of the treatment train. The organic strength in sewage has been considered too low for economical treatment by anaerobic systems (although this perception may well change in the future).

As highlighted in Sec. 5.4, anaerobic processes can also be classified as suspended growth, biofilm, and hybrid systems. Examples from each of these types are as follows:

(1) Anaerobic lagoon
Since all types of lagoons have large area requirements they are rarely used for industrial wastewater treatment in urban areas. Contrary to common perception

anaerobic lagoons do not necessarily have a serious odor problem although some odor can be expected. The exception to this would be when the wastewater is high in sulphates. Given the space requirements and the possibility of odor issue, anaerobic lagoons are usually found away from urban areas and this means they are often associated with agricultural or agro-industrial wastewaters. In Asia, anaerobic lagoons are frequently used to treat palm oil mill effluent (POME), a wastewater which has very high organic strength. Typically lagoons (anaerobic and other types) are of earthen construction. The excavated material is used to construct the bunds and, where it is available locally, the lagoon would be lined with clay. If the latter is not available, synthetic liners would be used. Ideally, anaerobic lagoons should be constructed deeper (4–5 m water depth compared to the facultative lagoon's 1.5 m) than the other types of lagoons and this is particularly so for wastewaters like POME because volume has to be allocated for storage and digestion of sludge derived from the wastewater's suspended material, and to reduce the surface area required relative to volume. Lagoons (but less so for anaerobic lagoons) are frequently operated as a series of cells. The benefits of such staging have been discussed in Sec. 5.3. Figure 5.5.1 shows the construction of two anaerobic lagoon cells in progress.

Anaerobic lagoons are operated without the covers usually associated with other types of anaerobic reactors. The exclusion of oxygen is achieved by the scum layer which would form on the mixed liquor surface. In wastewaters like POME and meat-processing wastewater, the O&G in the wastewater helps to form the scum layer. The need for this scum layer precludes the widespread application of anaerobic lagoons in series as the downstream cells may have difficulty forming an adequate scum cover as the O&G is progressively removed. Two anaerobic lagoon cells in series would typically be the configuration encountered. Figure 10.3.1 shows an anaerobic lagoon which has been treating POME for some time and the scum layer which has formed over time is clearly visible. This scum layer may sometimes even look misleadingly solid and is therefore potentially dangerous to those not familiar with such operations. Since anaerobic lagoons are perceived to be relatively trouble free once they have been successfully started up, their maintenance can be poor. As evident from Fig. 10.3.1 (where the bunds are no longer evident), erosion of the bunds is a problem and this is common in tropical areas where rainfall can be heavy over short periods. Bund erosion results in the eroded material entering the lagoon and hence reducing the lagoon's design hydraulic capacity. The eroded bunds would not keep surface runoff away from the lagoon. Consequently, the lagoon's design hydraulic capacity may be exceeded by a wide margin during such rainfall episodes leading to washout and possibly process instability and eventual failure. Inlet and outlet structures are located at

Fig. 5.5.1. Two anaerobic lagoon cells under construction. Note that the lagoon cells had not been lined nor had the bunds been fully constructed. The reinforced concrete inlet and outlet works had yet to be constructed.

the ends of the lagoon along the longitudinal line. In the absence of windbreaks, a series of lagoon cells (and indeed even the individual cell), where possible, should not be oriented such that the direction of wastewater flow through the lagoons is the same as that of the prevailing winds at the site. This is to reduce the incidence of shortcircuiting across the surface. For a similar reason, the inflow of wastewater would not be at the surface of a lagoon but located near to the bottom. This arrangement also brings the incoming wastewater into contact with the sludge blanket accumulated on the lagoon's bottom. Lagoon overflow would not be from the surface but a short distance beneath the scum layer. This is to avoid drawing out the scum. Anaerobic lagoons are not mixed except for the mixing caused by the release of biogas and the inlet to outlet flow pattern. In the application of anaerobic lagoons, the biogas generated is unlikely to be collected but is allowed to escape through the scum layer into the atmosphere.

Anaerobic lagoons, like any other anaerobic process, are sensitive to pH. During start-up or when a lagoon has received a shock load of high strength wastewater, pH may decline because methanogenesis may not be able to cope with the increased acidogenesis. Lime, soda, or soda ash would then have to be added into the lagoon to adjust the pH upwards. This pH adjustment may have to be continued until such time sufficient numbers of methanogens have accumulated in the lagoon to remove the volatile fatty acids formed by the acidogens. Typical loadings imposed on anaerobic lagoons range from 300–400 g $BOD_5$ $m^3$ lagoon volume $d^{-1}$. Underloading an anaerobic lagoon, perhaps from overly effective pretreatment, can also cause problems. The consequent low BOD loading and O&G content may result in anaerobic conditions not developing throughout the depth of the lagoon. This results, in effect, in a facultative lagoon and such facultative lagoons have been known to cause serious odor problems.

If practiced, desludging anaerobic lagoons can be a relatively frequent activity compared to other lagoon systems like the facultative and aerobic lagoons. This is because the anaerobic lagoon typically receives the strongest influent in the treatment train. Aside from anaerobiosis, suspended solids in the wastewater are also removed by sedimentation. Desludging may be achieved using draglines although handling the wet sludge, which like a slurry, can be difficult. A more common practice, especially for agricultural or agro-industrial wastewaters like POME, is to divert the wastewater to another newly constructed lagoon and to allow the first lagoon to dry out. An anaerobic lagoon is desludged or closed for drying when it is determined to be about half filled with sludge.

(2) Anaerobic digester
Conventional complete mix digesters are sometimes used to treat very strong wastewaters. These are operated as once-through reactors and their hydraulic retention times (HRT) are equal to the cell residence times (CRT). The limiting HRT is reached when the methanogens (the rate limiting component in the anaerobic consortium) are washed out faster than they can reproduce to replace those lost. HRTs are typically 30 days or longer. The wastewater may be fed into the digester continuously or intermittently and effluent is simultaneously withdrawn. Some degree of stratification does occur in the digester with a scum layer at the top followed by the supernatant and actively digesting sludge layers, and at the bottom the digested sludge layer. Mixing can be performed by mechanical mixers or gas injection through gas lances and draft tubes. Field experience with mechanical mixers, in the context of industrial wastewater treatment in Asia, has on many occasions not been as satisfactory compared to gas mixing in terms mechanical reliability and energy consumption considerations. The mixing, aside from enhancing contact between substrates and the microbial population, also helps to break up the scum layer. Unlike anaerobic lagoons where the scum layer is a desirable feature, the scum layer in digesters (and the remaining types of anaerobic

reactor configurations discussed in this section) is undesirable as it interferes with reactor operation and should be destroyed.

The POME identified in the preceding section on lagoons has also been frequently treated with anaerobic digesters and these are usually designed as high-rate digesters. In such digesters the CRT is separated from the HRT and this is possible by having a settling tank after the anaerobic digester and operated in a manner analogous to the activated sludge process in aerobic systems. A key difference is the insertion of a degasifying unit between the anaerobic digester and settling tank. This unit, operated under vacuum, is necessary to strip the anaerobic liquor of biogas before it reaches the settling tank, failing which liquid-solids separation and solids settlement would be poor. Another noteworthy difference is such settling tanks are also designed to provide a higher degree of thickening than would be usual in the clarifiers in the activated sludge systems. These settling tanks receive liquor with 10 000–30 000 mg L$^{-1}$ SS. Digesters operate with covers which are either fixed or floating. Where a cluster of digesters is operated (Fig. 5.5.2), the digesters may in the interest of costs be installed with fixed covers. The gas collected from the digesters is led to a storage tank with a floating cover. This storage tank would be filled with water to provide the seal and the

Fig. 5.5.2. A cluster of three anaerobic digesters used at a sugar mill. Such digesters can also be found at palm oil mills.

cover is fitted with counterweights so that it may move smoothly in the vertical direction as gas is drawn into or out of the storage tank. It is important to avoid the creation of negative pressures (relative to atmospheric pressure) as gas is flared or withdrawn from storage for use in boilers and gas turbines for steam and electricity generation. Such negative pressures can result in air being drawn into the digesters. Oxygen in the air can inhibit the biological process and possibly also create an explosive air-methane mixture.

Anaerobic digesters can be difficult to start-up because of the ease with which it is to upset the balance between the acidogens and methanogens. Ideally the seed biomass should come from an anaerobic digester treating a similar wastewater but in the absence of this, sludge from an anaerobic digester at a sewage treatment plant has been used. In rural areas where even the latter is not available, cow dung has been used as the seeding material. The usual procedure is to fill the digester with the wastewater to be treated and then to add in 20–50 $m^3$ of active and related anaerobic sludge. The digester is then fed with progressively increasing amounts (in a stepwise manner) of the wastewater and should be operational in 4–6 weeks. During the start-up period, lime or some other alkali would be added to maintain pH. Failure to do this could result in a sharp pH drop which is very detrimental to the development of the methanogens and hence delay completion of start-up. Erratic addition of alkali, leading to sharp pH swings can be just as detrimental. Monitoring digester performance would include measurement of gas production against organic loading and determination of gas quality in terms of carbon dioxide and methane. Initially carbon dioxide content can be higher than methane but should decline as the process stabilizes and the methane content is eventually 55–70%. A broad range of gas yields, 250–900 L $kg^{-1}$ $BOD_5$ removed, has been encountered onsite. Much would depend on the wastewater being treated and how well the digester has been operated. Typical digester loadings range from 1.0–2.5 kg volatile solids $m^{-3}$ $d^{-1}$ but these can be much higher for specific types of wastewater. For example in POME treatment a 5 kg volatile solids $m^{-3}$ $d^{-1}$ loading is not unusual.

(3) Upflow anaerobic sludge blanket (UASB) reactor

The preceding descriptions of the anaerobic lagoon and digester have identified the presence of a sludge blanket. The formation of such a layer of sludge in anaerobic reactors is not as difficult as expected. This is because anaerobic sludge inherently flocculates and hence settles well. This, of course, is on the proviso certain physical and chemical characteristics are present in the operating environment. The UASB exploits this feature of the anaerobic sludge. A UASB reactor operates with three zones — the sludge bed, sludge blanket, and the gas separation and settling zone. In terms of reactor configuration, the UASB reactor has three distinct parts — The digestion compartment, gas-solids separator, and the settler. The digestion compartment would be below the gas-solids separator while the settler would be above. The gas-solids separator separates the biogas bubbles from the solids moving up from the sludge blanket. This is important if the solids are to be directed downwards again, back towards the digestion compartment by the settler. The performance of the gas-solids separator becomes particularly important when the loading rate on the UASB reactor is high. A UASB reactor also has a feed distribution system which attempts to evenly distribute the incoming wastewater over the bottom of the reactor. This helps reduce the incidence of channeling. The number of feed inlets installed to achieve this can range from 1 for every 1 m$^2$ to 1 for every 5 m$^2$ of reactor base area. Much would depend on the loading rates — the higher the loading rate the fewer feed inlets because the biogas generated helps in mixing and hence reduces the risk of channeling in the sludge bed and blanket. Treatment performance can decline quickly should such channeling occur. The biomass concentration in the sludge bed could be 40–70 g VSS L$^{-1}$ while the sludge blanket above it has concentrations of 20–40 g VSS L$^{-1}$.

Where possible, a new UASB reactor should be seeded with sludge drawn from another UASB reactor. This ensures the seed material is granular in nature and of high specific activity. During start-up some degree of wash out is desirable so as to remove poorly settleable material and dispersed filamentous microbes. This is to ensure the heavier and hence more settleable sludge is accumulated. To achieve this, and if the wastewater is too strong, effluent recycle or dilution may be necessary. Organic loadings should be increased stepwise gradually so as not to create conditions which would result in a loss of balance between the acidogens and methanogens. In addition to this, it is also desirable to enhance the growth and accumulation of *methanothrix*-type bacteria over *methanosarcina*-type bacteria. The latter is undesirable because of their relatively low activity when acetate concentrations are low which would be the case when the reactor has reached steady-state. Consequently, during start-up, it is important to ensure that conditions favoring *methanosarcina*-type bacteria growth like acetate concentrations above 500 mg L$^{-1}$ and pH below 6.5 are avoided. The *methanothrix*-type bacteria are rod-shaped and would agglomerate into spherical granules of about 1–3 mm diameter. The formation of granules in some wastewaters appears to be enhanced when Mg$^{2+}$ or Ca$^{2+}$ is supplemented. However, this practice may not be suitable if continued because of the risks of scaling. Following start-up, loadings on UASB reactors can range from 5–20 kg COD m$^{-3}$ d$^{-1}$. UASBs can be sensitive to hydraulic surges as this may cause loss of biomass. The biomass may also be displaced from the UASB if the wastewater it is treating contained substantial quantities of suspended particles. The latter may accumulate in the reactor, hence reducing its effective volume.

(4) Anaerobic sequencing batch reactor (anaerobic SBR)
The anaerobic SBR is very similar in concept to the aerobic SBR (which shall be discussed in a later section). Among the major differences are, of course, the absence of aeration and the presence of an air-tight cover in the anaerobic SBR. The key difference between the anaerobic SBR and the other anaerobic processes discussed in this chapter is that it is a batch (or cyclic) process. There are typically five phases in each cycle of reactor operation — FILL, REACT, SETTLE, DECANT, and IDLE. During FILL, wastewater is received by the reactor and the latter begins with the sludge held over from the previous cycle occupying an assigned portion of the reactor's working volume. FILL ends either when the reactor has reached the maximum water volume it can work with or when a maximum time for FILL has elapsed. REACT then begins and this phase is likely to be the only phase in the entire cycle where mechanical mixing (other than that inherent with FILL and biogas generation during FILL and REACT) is initiated. The mechanical mixing can be performed by pumping mixed liquor from the upper portion of the reactor and returning this via a distribution box to the base of the reactor through a number of inlets. The criteria for the number of such inlets are similar to that for the UASB. At the end of REACT, the mechanical mixing is stopped and SETTLE begins. Liquid-solids separation takes place under relatively quiescent conditions and the sludge should form a distinct blanket with the sludge solids — supernatant interface below the decant device. With DECANT either an automatic decant valve is opened or a decant pump is activated. Typically the decant pipe is fitted with an anti-vortex device to minimize vortex formation in the reactor during DECANT and hence re-suspend the settled sludge particles. The anaerobically treated and clarified wastewater is discharged until a preset water level in the reactor is reached, following which the valve or pump would be closed or deactivated. Once the DECANT phase has been completed, the reactor is in IDLE phase and is ready to receive its next batch of wastewater. Figure 8.3.3 shows an example of the anaerobic SBR. The cover of the anaerobic SBR is an important feature as it has to be capable of considerable vertical travel when the reactor is charged with wastewater (upwards movement) and when it is decanted (downward movement). Failure to do this results in negative pressures and air being drawn into the reactor headspace. The anaerobic SBR example shown has a "red-mud" rubber membrane cover (Fig. 8.3.4) and this collapses during DECANT and fills up (like a balloon) during FILL. It expands further during much of REACT as biogas is generated. Desirable membrane cover material

characteristics include UV stability and low gas permeability. This is to ensure that the membrane cover does not easily deteriorate when used under exposed conditions, resulting in gaseous escape. The anaerobic SBR reactor typically operates with a maximum side water depth of 4–5 m and have a rectangular vessel configuration. The decant device would be located at one of the narrow ends with the decant pipe punching through the reactor's wall if the supernatant is drained under gravity instead of being pumped.

The anaerobic SBR, being a batch process, is a non-steady state process. The substrates profile in the reactor is such that it starts low (at discharge concentrations) at the beginning of FILL, builds up during FILL, and then begins to decline to the discharge concentration during REACT. During SETTLE and DECANT the residual substrates concentration should be at the discharge concentration. Such a substrates profile is important to the successful operation of the anaerobic SBR. The high food to microbe ratio during FILL and the earlier part of REACT would result in high microbial activity and biogas formation. The high biogas generation also helps in mixing the reactor's contents. This is supported by the mechanical mixing during REACT which also helps to strip the gas bubbles from the sludge particles, releasing these into the headspace. Mechanical mixing is particularly important towards the end of REACT. The food to microbe ratio should be very low towards the end of REACT, and during SETTLE and DECANT. This means microbial activity has already declined very significantly as would have biogas production. The decline in biogas production is a key contributor to effective liquid-solids separation and sludge blanket formation prior to DECANT. The solids content in the sludge blanket can be expected to be upwards of $20\,g\,VSS\,L^{-1}$. Scum formation and accumulation is a possible difficulty which can result in poorer effluent quality. Figure 5.5.3 shows the presence of such a scum layer in the anaerobic SBR after it has been taken offline for maintenance. However, during regular reactor operation scum has not been noted to be an issue, possibly because the mechanical mixing reduces its formation in the first instance and also breaks up the scum as it forms.

While the anaerobic SBR would benefit from granular sludge, there are unlikely to be sufficient numbers of them in operation treating various types of wastewaters to provide the seed material at the time of this book's preparation. Consequently anaerobic SBRs are likely to be seeded with flocculant sludge from anaerobic digesters. Over time granular sludge about 1 mm in diameter may form but field experience has not indicated these granules would completely replace flocculant sludge. Instead the granules appear in large numbers mixed with the flocculant sludge. It may be argued the operating mode of the anaerobic SBR favors selection of heavier well-settling sludge particles since the DECANT phase

Fig. 5.5.3. Uncovered anaerobic SBR in the foreground showing accumulation of a scum layer. The dismantled membrane cover may be seen on the LHS.

tends to wash-out poorly settling flocs and dispersed organisms although the wash-out mechanism may not be aggressive enough to remove the flocculant sludge while retaining only the granules. The use of polymers during start-up to enhance flocculation has been attempted and cationic polymers have been effective. These polymers may have been effective because they form a bridge between negatively charged bacteria cells through electrostatic charge attraction and they also create large biomass aggregates via sweep-floc mechanisms. As in the UASB during start-up, substrate loading has to be progressively increased. Failing this an inappropriate mix of *Syntrophomonas spp., Methanosarcina spp.,* and *Methanothrix spp.* may result. Following start-up, reactor organic substrate loadings at 6–9 kg COD m$^{-3}$ d$^{-1}$ can be beneficial to granule formation and growth.

Anaerobic SBRs have been successfully operated in sequence with the aerobic SBRs because they share similar control protocols. This makes control software preparation for the programmable logic controller (PLC) a relatively simple matter. The anaerobic SBR would provide pretreatment to reduce organic strength while the aerobic SBR provided the polishing.

(5) Anaerobic filter
The anaerobic filter, unlike the UASB and anaerobic SBR, depends on a bed of packing material (Fig. 5.4.3) to reduce washout of biomass from the reactor at short HRTs. The wastewater may pass through the anaerobic filter in the upflow or downflow mode although the upflow mode appears to be more common. Even distribution of the wastewater across the cross-sectional face of the support medium is important to reduce risk of channeling and hence a deterioration in process performance. This can be achieved by feeding through evenly spaced feed inlets and placing the support medium over a flow dispersion plate. During reactor operation, the bed of support medium is completely submerged. In the upflow filter, the wastewater rises through the support medium and exits the anaerobic filter via the overflow box. The overflow box is typically provided with a water seal to prevent ingress of air. Anaerobic filters can have an overall height of 5–7 m and are often cylindrical in shape (although square cross-sections are not unknown). Various packing material have been employed but moulded plastic shapes have been commonly used. High bed porosity and surface area are desirable features. The plastic material and high porosity go towards reducing the weight of the support medium bed and this helps to reduce construction costs. The high surface area is conducive to the formation of biofilms and hence immobilization of bacteria. Specific surface areas of loose pack plastic media can be about 100 m$^2$ m$^{-3}$ with a void fraction of about 95%. The density of the support medium is usually much less than 100 kg m$^{-3}$. Gravel is rarely used in anaerobic filters treating industrial wastewaters. This is because it would not be able to provide the high specific surface area, void fraction, and weight desired. In cylinder-shaped reactors, the base may be shaped as a shallow cone. This is to facilitate sludge collection and removal. Desludging typically takes place once or twice a year depending on the nature of the wastewater treated.

Depending on the support medium used, biomass concentration in the anaerobic filter can be expected to be upwards of 20 g VSS L$^{-1}$. Examples of bacteria which may be immobilized in the biofilm are *Syntrophomonas spp.,*

*Methanococcus spp.*, and *Methanothrix spp.* Determining the quantity of biomass in an anaerobic filter can be somewhat more complicated than the preceding reactor types because biomass can be retained by a combination of two mechanisms. The first is obviously by means of biofilm formation and bacteria are immobilized on the surface of the support medium. However, flocculant bacteria is also retained in the voids formed by the loose packing of the moulded plastic shapes. There is obviously a limit to the amount of biomass which can be held in these voids before filter bed clogging occurs. Clogging can become particularly serious if the wastewater contained particulate matter and this is especially so for stationary bed anaerobic filters. Sometimes when the anaerobic filter is deemed to be a suitable treatment option if it had not been for its relative intolerance to particulate matter in the feed, a moving bed filter may be attempted. A reactor configuration which incorporates the true moving bed as in a fluidized bed is not common in practice. The effect of a moving bed may be achieved to some extent by the use of flexible fibers as the support medium as shown in Fig. 5.4.3. This arrangement is much more tolerant to particulate matter in the feed wastewater and the anaerobic filter shown in Fig. 5.5.4 received wastewater with 200–300 mg L$^{-1}$ SS. In this configuration the biomass is primarily retained in the reactor by immobilization on the fibers.

Fig. 5.5.4. IWTP comprising of a SBR and anaerobic filter (RHS cylindrical reactor) for fructose wastewater.

Biogas generation within the anaerobic filter provides some degree of mixing but this is usually insufficient in achieving mixed-flow conditions. The flow regime would be closer to the plug-flow condition. It has been noted the latter can lead to local accumulation of VFAs in the reactor and progressive process failure. In addition to this, such anaerobic filters have also been noted to be more susceptible to the effects of inhibitory substances in the wastewater. Effluent recirculation has been found to alleviate such difficulties. The increased flow through the filter bed also released entrapped biogas and scum which would otherwise have reduced the effective volume of the support medium bed. Effective recirculation flows have ranged from 0.5 to 4 times bed volume per hour. The appropriate recirculation rate depends on the type of support medium used and the wastewater undergoing treatment. Excessively high recirculation rates are not desirable as the biofilms may be sheared away from the support medium and a sufficiently thick biofilm cannot then be formed. However, by the same token increasing the recirculation flow periodically for short durations has been found to be effective at reducing the incidence of clogging. This may be achieved by putting the standby recirculation pump into service during such periods. "Pulsing" the flow has been particularly useful when treating high strength wastewater and the anaerobic filter is heavily loaded. Even if the wastewater has a low particulate content, the large amount of soluble substrates present would result in a larger yield of biomass. The latter would need to be removed periodically to reduce the incidence of clogging. "Backwashing" or "pulsing" with the recirculation pumps arrangement has been found effective. With recirculation to enhance mixing within the anaerobic filter, it has been found to be tolerant to organic shock loadings, pH variations, and inadvertent introductions of inhibitory substances. The anaerobic filter has been noted to be capable of withstanding considerable surges in hydraulic loads before biomass washout occurs. In this, the anaerobic filter with recirculation is almost always more robust than the preceding anaerobic reactor types.

(6) Hybrid anaerobic reactor
The hybrid anaerobic reactor combines the sludge blanket with the anaerobic biofilm. Hybrid reactors are constructed with 3 major zones — the sludge thickening zone in the bottom hopper, the sludge blanket zone in the middle, and the biofilm zone above the sludge blanket. The biofilm support medium can be of shaped plastic blocks which allow a pattern of channels to be formed. These would, aside from creating surface area for biofilm adhesion, also assist in gas-solids disengagement. Such reactors may be started using flocculant anaerobic sludge to develop the sludge blanket. The biofilm would develop from biomass

Fig. 5.5.5. IWTP with hybrid anaerobic reactors (the pair of cylindrical reactors on the LHS) and activated sludge process for rubber thread wastewater.

washed up from the sludge blanket. Purpose-built hybrid anaerobic reactors are still rare in the industry. Figure 5.5.5 shows a pair of these used to treat wastewater from the rubber industry.

## 5.6. Aerobic Processes

In industrial wastewater treatment, aerobic processes can follow anaerobic processes to provide the additional treatment to improve the quality of the pretreated effluents to discharge limits or, where the wastewater organic strength had not been so high in the first place, to provide the only biological treatment needed to produce the treated effluent quality. Many of the aerobic processes encountered in industrial wastewater treatment are of the suspended growth types although biofilm types are also known. Since these are aerobic processes, oxygen would have to be supplied. In most of these processes, oxygen transfer into the reactors' mixed liquor is achieved by moving air from the atmosphere into it. The use of pure oxygen or oxygen enriched air is rare in Asia.

Design parameters which are important to suspended growth aerobic processes include the flow regime, cell residence time, F:M ratio, MLVSS concentration, aeration period, cell yield, and recirculation ratio. Many of the suspended growth aerobic processes which can be encountered are variants of the activated sludge process. While they may be variants, their design parameters can show substantial differences. Some of the common variants are the conventional, step aeration, contact stabilization, complete-mix, extended aeration, pure oxygen, and sequencing batch reactors. With the exception of the sequencing batch reactor, the rest are continuous flow processes. The sequencing batch reactor is a cyclic (sometimes referred to as an intermittent) process.

(1) Activated sludge processes
Although there are numerous variants, the common types encountered in industrial wastewater treatment include the conventional, complete-mix, and extended aeration. The conventional process has the influent and returned sludge entering the aeration basin at its head and the mixed liquor leaving it at the opposite end. Mixing within the basin is caused by the aeration. The flow regime is plug flow. The complete-mix process has the influent and returned sludge entering the aeration basin along its length and the mixed liquor then flowing across the basin to the effluent channel. This arrangement, with the aeration, creates complete-mix conditions within the basin. The extended aeration process typically operates with aeration basins which are larger than the other two variants to treat a similar amount of wastewater. This means that the pollutant loading imposed on it in terms of per unit reactor volume or unit of MLVSS (ie. F:M ratio) can be much lower. The flow regime can be either complete-mix or plug flow.

A modification of the extended aeration process is the oxidation ditch. This has a relatively long basin with a central dividing wall creating two channels in the shape of a race track (Fig. 5.6.1). Aeration in these is provided by rotor brushes — one at each end of the basin. These rotor brushes not only aerate the mixed liquor but also transfer a horizontal vector to it so that a particle of water upon entering the oxidation ditch would move through a channel and into the next in a plug flow manner. The oxidation ditch would be a rectangular basin with rounded ends. Being a plug flow reactor, it is vessel shape sensitive. The conventional activated sludge process is also housed in rectangular vessels while the complete-mix and extended aeration (complete-mix regime) can have circular or square vessels. Square and rectangular shapes are more typical among vessels used in industrial wastewater treatment unless the vessels are of steel construction. In the latter case a circular tank construction is often adopted for structural strength reasons. Square (and rectangular) vessels can be easier to arrange in constrained spaces and there can be savings in construction costs if the vessels shared walls. With the exception of oxidation ditches which are almost invariably constructed of reinforced concrete, vessels for the other variants may be of reinforced

Fig. 5.6.1. An oxidation ditch at a palm oil refinery. Note the central wall which divided the basin into two halves and thereby forming the "race track" configuration.

concrete or steel construction. These vessels have side water depths of 3–6 m and freeboards of about 0.5 m. Water depths of less than 3 m may result in oxygen transfer efficiencies which are lower than acceptable values.

Table 5.6.1 compares the values of some of the design parameters used for these three processes. Each of the five parameters shown in the table can affect the size of the aeration vessel. As CRT, recirculation ratio and aeration period increase, vessel size increases but the reverse is true as MLVSS and F:M ratio increased. The designer would have to choose an appropriate combination of these parameters to not only achieve the specific process and performance targeted but also to

Table 5.6.1. Values of design parameters for activated sludge process variants.

| Process | CRT, d | F:M, $d^{-1}$ | MLVSS, mg $L^{-1}$ | Recirculation ratio | Aeration period, h |
|---|---|---|---|---|---|
| Conventional | 5–20 | 0.2–0.4 | 1200–2500 | 0.3–0.5 | 4–8 |
| Complete-mix | 5–20 | 0.2–0.6 | 2500–4500 | 0.3–1.0 | 3–5 |
| Extended aeration | 20–30 | 0.05–0.2 | 2500–4500 | 0.5–2.0 | 18–36 |

Note: The F:M ratios have been calculated in terms of $BOD_5$ and MLVSS.

accommodate future changes. For example, vessel size can be reduced by increasing the MLVSS and/or F:M ratio but if a system is designed to operate with high values from the onset, there would be little opportunity to increase these values to accommodate increases in pollutant loads generated at the factory. The latter occurs frequently, and sooner than expected, after a factory has entered production and its products find increasing success in the market. Production would then need to be expanded and this can be achieved very quickly in the first instance without adding production lines but by increasing the number of shifts (eg. from 1 shift to 2 shifts $d^{-1}$). Given the frequency of such occurrences, it may be prudent to design with the lower values shown on Table 5.6.1.

Given the three activated sludge variants shown in Table 5.6.1 have a continuous flow regime and many factories operate on a shift basis, there is therefore a problem with supplying wastewater continuously to the activated sludge process unless holding capacity has been included upstream of it. In the absence of such holding capacity the intermittent flows into the aeration basin can lead to process instability and fluctuating treated effluent quality. In situations where complying with discharge limits is on the basis of average values, the treatment facility may still be able to perform satisfactorily. This cannot be the case where absolute limits on pollutant concentrations in the effluent apply.

In each of these three processes, sludge return from the secondary clarifier to the aeration basin is an important operation. This is done by drawing sludge from the hopper of the secondary clarifier at predetermined recirculation rates. The recirculation ratios in Table 5.6.1, which are calculated by comparing the returned sludge flow rate against the influent flow rate, provide the range of values encountered. Failure to maintain adequate sludge return would deplete the aeration basin's MLVSS and this has the effect of increasing the F:M ratio. A consequence of this would be deteriorating treated effluent quality. It would be prudent to allow for some flexibility in the recirculation ratios when specifying the sludge return pump capacities.

The decision on which process variant to use on a particular wastewater should depend on the latter's characteristics. A key consideration in industrial wastewater treatment is the presence or absence of potentially inhibitory components. Should the latter be present and these are allowed entry into the aeration vessel, then rapid dispersion of such components to lower concentrations throughout the vessel is desirable. In such an application the complete-mix reactor has an advantage. If the wastewater contained Amm-N and nitrification is required, any of the three process variants could have achieved it although the extended aeration process has an advantage given its longer CRT (and hence higher accumulation of nitrifiers). However, if nitrification and denitrification are required then the plug flow reactor (eg. the oxidation ditch) has an advantage since it would be simpler to create oxic (for nitrification) and anoxic zones (for denitrification) in such reactors.

Aeration can be achieved by any one of three commonly used methods. In the urban setting, aeration is usually with diffused air. In this arrangement, a blower forces air through diffusers placed on a grid of pipes anchored on the base of the aeration basin. These diffusers may be of the coarse or fine air type. The latter has become increasingly common since these allow for higher oxygen transfer efficiencies (of up to 36% transfer efficiency compared to the 8% from coarse air diffusers) and hence better energy efficiency. In the past these fine air diffusers were ceramic diffusers but in the last 20 years flexible membrane diffusers have become increasingly common. Field experience suggests that these have lower maintenance requirements since they suffer less from clogging. Membrane diffusers may either be of the circular disc type (Fig. 8.3.6) or the flexible tubular type (Fig. 5.6.2). Both have been found effective in terms of oxygen transfer although the tubular type may be somewhat easier to lift out of the aeration basin for cleaning if the aeration system's operating protocol requires this. The well designed diffuser grid can be expected to provide the most complete mixing in an aeration

Fig. 5.6.2. Tubular diffusers installed on the bottom of an aeration vessel.

Fig. 5.6.3. Jet aerators in a reactor at an IWTP for a milk canning factory. These are normally submerged and are shown exposed in the picture because the reactor has been drawn down for maintenance.

basin without the "dead zones" which may be present if the jet or surface aerator is used (Fig. 5.6.3).

To reduce the noise nuisance from blowers supplying air to the diffusers, these are located either in sound-proof enclosures or blowerhouses provided with noise attenuation fittings. Given the large amount of air which has to be injected into the mixed liquor in an aeration basin, aerosols can be a concern. Aeration basins have been covered to mitigate aerosol concerns. Jet aerators may operate without blowers. These mix compressed air and liquid within the aerator before releasing the mixture into the aeration basin. The rising plume of fine air bubbles produces mixing and oxygen transfer. Where air requirements are relatively low, the air may be drawn into the aerator by an aspirator pipe arrangement leading from the atmosphere to the aerator. This would be inadequate if the air requirement is high and a blower then supplies the air. Jet aerators are usually almost as energy efficient as the fine air diffusion systems and if operated without blowers are the quietest of the three aeration options. They do, however, suffer from nozzle or venturi clogging if the wastewater had not been pre-screened well. They also tend to work better in the deeper aeration basins. The third common aeration method is surface aeration using either high or low speed surface aerators. Surface aerators are either mounted on fixed structures such as platforms or bridges (Fig. 9.3.3), or on floats (Fig. 10.3.2). Since the motors are mounted above the impeller and are therefore above water, noise can often be an issue. The noise from high speed aerators can be particularly annoying to neighbors. Surface aerators are common in agricultural and agro-industrial wastewater treatment and are associated with aerated lagoons (the conventional activated sludge process housed in a large earthen basin). Surface aerators are usually associated with water depths of less than 4 m since mixing would probably be inadequate at greater depths. The low speed surface aerator is more expensive than the high speed aerator but the former provides better mixing. Surface aerators have been used to replace the brush rotors used in oxidation ditches because they are cheaper. Where surface aerators are used with earthen lagoons, a concrete scour pad should be placed on the base of the lagoon directly beneath the aerator. This is to prevent erosion of the base. Similarly the banks of the earthen lagoon would have to be lined (eg. with concrete slabs) to protect these from erosion caused by the "wave" action resulting from surface aerator operation. In rectangular basins, more than one aerator is used to reduce the incidence of poorly mixed zones.

Raised mixed liquor temperature is an issue which can be associated with diffused air aeration in deep vessels. For example in vessels with side water depths of 5 m, mixed liquor temperature can be 36–40°C by late afternoon in the tropics (compared with ambient water temperatures of 26–30°C). This increase in temperature is the result of the air forced into the mixed liquor by the blower. The deeper the vessel, the higher the temperature of the air as the blower moves it into the aeration grid and through the fine air diffusers against the water column head. The higher temperatures reduce oxygen solubility in water while increasing the biological degradation rates.

Foaming can occur in aerated processes because of the presence of detergents and protein substances associated with the wastewater, or polymeric substances released by the microbial population during treatment of the wastewater. In the last case, this may have been due to operating F:M ratios which are too low. These are conditions which can occur during plant start-up. Foaming is undesirable because it results in uncontrolled biomass loss from the aeration basin. In extreme cases the foam may even overflow the aeration vessel (Fig. 8.4.1). Foaming also raises aesthetic and health concerns as the foam can be carried away from the treatment facility by wind. Where there are substances associated with the wastewater which are suspected as being capable of causing foaming, then these should be removed before the aeration basin. If such removal is not possible then the aeration basin can be operated with an anti-foam dosing system. A silicone-based anti-foam agent (non inhibitory to the microbial culture) would be able to disperse the foam. Dosages required vary with the application and is best determined on site. Alternatively sprays can be installed around the aeration vessel (Fig. 3.4.1). The water sprays break the foam up before excessive amounts accumulate.

Bulking sludge can be a problem with the activated sludge process and this can afflict all its variants. Sludge bulking occurs when filamentous bacteria multiplies and begins to dominate the microbial consortium (Fig. 7.4.3). When this occurs, sludge compaction during settling deteriorates and sludge with lower solids content would then be present in the clarifier hopper. Since the recirculation rate to return sludge to the aeration vessel is estimated on the basis of a given solids content, returning the same volume but with a lower solids content results in progressive depletion of MLSS in the aeration vessel. An indication of the proliferation of such microbes can be provided by the sludge volume index (SVI) test. SVI values of 80 to 120 would be desirable. Values of 150 or more would suggest a bulking sludge and values of more than 200 would be badly bulking. The latter would not only cause difficulties with the recirculation ratios and hence adequate sludge return but also adversely impact on the effectiveness of the sludge dewatering stage. Bulking sludges do not dewater well and hence result in a wet sludge cake. It should be noted MLSS with very low SVIs (ie. less than 80) may not necessarily be desirable either as the absence of filamentous growth means the absence of a matrix for the sludge flocs to form on. This may result in dispersed growth which does not settle well and hence settled effluents which are turbid.

(2) Sequencing batch reactor (SBR) process
The SBR process became more commonly applied in Asia from the mid-1980s onwards as an alternative to the more commonly encountered continuous flow systems. It is the only commonly applied activated sludge variant which is designed to operate in a cyclic or intermittent mode. Because of the latter, the operation of SBRs can be matched with the shift nature of factory operations more easily than continuous flow systems. The differences between treatment trains incorporating the continuous flow activated sludge processes and the SBR begin from the aeration vessel onwards. Typically the continuous flow activated sludge process operates with aeration vessels and secondary clarifiers. There would be sludge return from the secondary clarifier to the aeration vessel. A SBR operates without the secondary clarifier and hence would also not have the sludge return from the latter.

The cyclic reactor variants which have been introduced over the last two decades may be broadly classified in terms of their feed and discharge patterns. The three common categories are the continuous feed-intermittent discharge, intermittent feed-intermittent discharge (ie. the SBR), and "reversing" flow. Of the three, the first two more closely resemble each other in terms of operating protocols and are more common in Asia. Reactors of the continuous feed type have a baffle wall at the head of the reactor where the wastewater first enters. This is to reduce the risk of untreated wastewater exiting the reactor during decant and so the reactor may be viewed as bi-chambered. Aside from this, the equipment used in the two variants is largely similar although arrangement details and equivalent numbers can differ. Of these equipment items, two clusters can make the SBR design very different from the systems discussed under activated sludge processes above. The SBR's decant system is probably unique to this type of bioreactors and this may take the form of a moving weir (mounted on an articulated arm — Fig. 3.6.4) or multiple fixed decant points. The former is more commonly found in the larger systems where the additional costs can be justified. Many industrial wastewater treatment plants, being smaller in terms of their hydraulic capacity, would opt for the multiple fixed point decanters (Fig. 5.6.4) and this decision is typically prompted by economic considerations. Such decanters do, however, require a better understanding of the process and biomass settling properties as the designer would need to estimate biomass settling velocities and the location of the sludge blanket-clarified effluent interface at the end of the settling phase. Inclusion of an anti-vortex device in the decanter, particularly the fixed point decanters, have been found useful in terms of avoiding vortex formation and hence resuspension of solids from the sludge blanket during decanting.

Apart from the decanters, the aeration systems of SBRs are usually also different, in terms of size, from those used in continuous flow systems. These differences have arisen because of the intermittent nature of aeration in SBRs. Because of the necessity to supply an equivalent amount of oxygen (ie. air) but over a shorter period in a day, the SBR aeration system is larger in terms of capacity compared to a continuous flow system treating an equivalent amount of wastewater. Coupled with this larger aeration capacity is the intermittent operation and together these can subject the diffuser grid to considerable stress. The design of the aeration piping grid would have to provide for even pressure distribution and better anchoring to the reactor's floor. Rapid and even pressure distribution throughout the piping grid can be achieved by designing a closed loop piping grid (Fig. 8.3.6) as opposed to the "fish-bone" piping configuration. It is also important to include a bleed line so that the aeration grid can be purged at regular intervals to remove water which can accumulate within it during periods in a cycle when aeration is cut-off. Field experience has indicated that flexible membranes diffusers have performed well with hardly any diffuser clogging problems. Where noise and spray are not issues, SBRs have used floating aerators held in place with

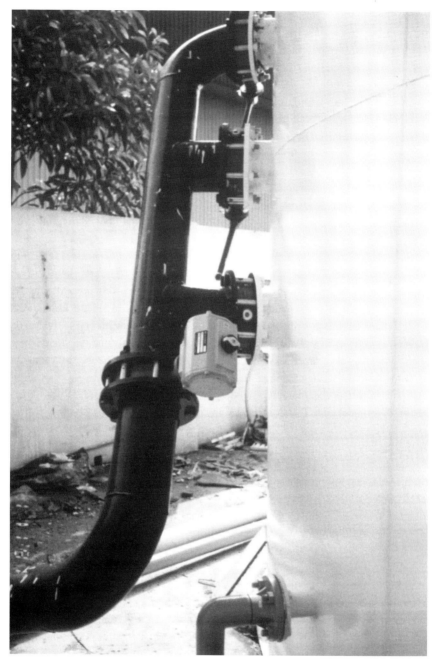

Fig. 5.6.4. Multiple fixed point decanter on a SBR which allows selection of any one of three preset decanted levels. The lowest and smaller pipe is for desludging.

guide cables mounted with counter-weights to facilitate smooth vertical movement of the aerators as water levels changed during FILL and DECANT. Such water level changes can range from 1–3 m depending on application while the retained mixed liquor volume at the end of DECANT would then correspondingly range from 4–2 m.

Each SBR can operate on the basis of a number of cycles in a day and each cycle can have 5 phases — FILL, REACT, SETTLE, DECANT, and IDLE. A SBR system can comprise one or more reactors. IWTPs typically have SBR systems with one to four reactors. The number of reactors selected depends on the volume of wastewater requiring treatment and the duration of its discharge over a 24 h period. SBRs have been designed with equivalent F:M ratios ranging from 0.05–0.6. This meant that the SBRs have simulated conventional activated sludge and extended aeration processes. MLSS concentration would typically not exceed 5000 mg L$^{-1}$ with values often within the range of 2000–3500 mg L$^{-1}$. Cycle times have usually been in the range of 6–12 h cycle$^{-1}$ although 24 h cycles are known. The FILL period may or may not be aerated although an unaerated FILL can be useful in terms of controlling sludge bulking and in denitrifying nitrified wastewater retained from the previous cycle. The SETTLE phase does not usually exceed 2 h. Long SETTLE periods are not desirable when dealing with well nitrified wastewaters. DECANT periods are usually within the range of 0.5–2 h and would, in part, depend on the weir loading recommended for the decanter selected. The IDLE period may even have zero time. Where time is allocated for this phase, then it can be used to accommodate more severe than usual flow fluctuations and, on occasion, for aerobic sludge digestion.

As with the continuous flow systems, SBRs can also be seriously affected by foaming caused, possibly, by *Norcadia spp.* Foaming in SBRs can frequently occur during the start-up phase when the equivalent F:M ratios are not as per design values. SBR operation can be very badly affected by foaming because the foam can "blind" the sensors used to control water levels, delay accumulation of an appropriate MLSS concentration because of uncontrolled biomass loss, and can cause serious deterioration in treated effluent quality as the foam enters the decanters during DECANT. Control of foaming (and bulking) can be achieved through an alternating "feast-famine" mode of operation. The intermittent feed nature of SBR operation inherently creates "feast" conditions at the beginning of a cycle and this would transit towards "famine" conditions towards the end of the REACT phase. This feature of the SBR can be manipulated to correct an existing condition or enhance control so as to reduce incidence of its occurrence. The intensity of the "feast-famine" condition can be varied by adjusting the ratio of the

FILL to REACT periods, the ratio of aerated to non-aerated periods, and the intensity of aeration during aerated periods. Known control strategies built around these methods suggest process control protocols which hinged on the growth kinetics of relevant groups of bacteria. Examination of biomass samples drawn from fullscale reactors has shown a reduction in filament numbers following introduction of an appropriate "feast-famine" strategy.

# CHAPTER 6

# THE INDUSTRIAL WASTEWATER TREATMENT PLANT — SLUDGE MANAGEMENT

## 6.1. Sludge Quantities

As discussed in Sec. 5.1, microbial populations in biological treatment processes convert part of the organic pollutants into new microbial cells. This means that the MLSS concentration in a reactor would increase over time unless the excess biomass is removed from the system. Excess biomass in the reactor is not a desirable condition since it may exert an oxygen demand which is beyond the capacity of the installed aeration system. Removal of this excess biomass can be achieved by desludging from the return sludge line leading from the secondary clarifier to the reactor in continuous flow systems. In SBRs, desludging can be performed towards the end of DECANT or during IDLE from the sludge sump constructed at a corner in the reactor. Where the electron acceptor is $O_2$ (eg. in aerobic systems), biomass yields, Y, can range from 0.3–0.8 kg VSS $kg^{-1}$ $BOD_5$ consumed depending on the nature of the substrate and the type of microbial consortium. Should the electron acceptor be other than $O_2$ (eg. in anaerobic systems) the yields can then be substantially lower than the values indicated. This microbial yield can be further reduced by manipulating the cell age of the microbial consortium. Younger cells will tend to have higher yields and vice-versa for older cells. The observed biomass yield values, $Y_{obs}$, take this into account and are therefore lower than the Y values. Table 5.6.1 shows typical CRTs ranging from 5 to 30 days for three activated sludge variants commonly used in industrial wastewater treatment. If minimizing excess biomass production is an objective, then the extended aeration process would have an edge over the conventional and complete-mix activated sludge processes. It should, however, be remembered that the actual amounts of sludge produced at a plant include not only the excess biomass but also any particulate matter which is inherent in the wastewater and which is not degraded. Of course, any other particulate matter which is generated during the treatment of wastewater other than by the biological process (eg. by chemical precipitation) and which may enter the biological reactor are also included. With this in view, the designer may well find the actual sludge yield values, $Y_{actual}$, for a particular reactor exceeding 1.0. Sludge wasting then becomes necessary not only to remove excess biomass but also to avoid accumulation of non-biological material which would, over time, reduce viability of the sludge within a reactor.

## 6.2. Sludge Thickening

While large IWTPs (eg. large animal husbandary farms, and pulp and paper mills) may include sludge thickening devices like the gravity thickener (Fig. 6.2.1), dissolved air flotator, and centrifuge, most plants serving individual factories are small enough in terms of the amount of sludge generated not to include such devices. Nevertheless some degree of thickening would be useful and can be expected to occur in the hoppers of secondary clarifiers and the sludge sumps within SBR vessels. While the MLSS concentration in the aeration vessels may range from $1500$–$6500\,\text{mg}\,\text{L}^{-1}$ depending on the type of process selected, the solids concentration of sludge drawn from the bottom hopper of the secondary clarifier or SBR sludge sump can range from $3000$–$12\,000\,\text{mg}\,\text{L}^{-1}$. Further sludge treatment such as digestion or dewatering can then performed on this.

Fig. 6.2.1. Gravity thickener at a large pig farm's wastewater treatment plant.

Withdrawing thickened sludge (eg. from a SBR's sludge sump) should not be at too high a rate, because a phenomenon, referred to as "rat holing" by plant operators, can occur. When "rat holing" occurs the sludge blanket in the immediate vicinity of the inlet to the sludge withdrawal pipe collapses inwards forming a hole above the pipe and leading to the surface of the sludge blanket. The sludge further away is, however, unable to move in quickly enough to "refill" this hole in the sludge blanket. The result of this phenomenon is that sludge with increasing water content is withdrawn following the initial discharge of higher solids content sludge. In extreme cases, the discharge eventually becomes fairly clear water.

## 6.3. Sludge Digestion

Anaerobic sludge digestion is rare in industrial wastewater treatment. As with thickening, the quantities of excess sludge produced typically does not justify its inclusion in a treatment train. Sludge holding capacity does, however, need to be provided to ensure, if necessary, that waste sludge can be stored pending dewatering. These open sludge holding tanks are aerated to reduce the formation of odors. These tanks have also been designed as aerobic digesters so that the organic matter in the sludge can be stabilized prior to dewatering. Where such tanks are designed only for the purpose of holding waste activated sludge, then their holding capacities would probably not exceed 3 days of excess sludge discharges from the aeration vessels. Capacities are considerably larger if these tanks are to serve as aerobic digesters and solids retention times can range from 10–20 days. The target, with such solids retention times, is then a 30–50% VSS reduction. Aeration can be with fine air diffusers and aeration rates of $1.2–2.0\,m^3$ air $m^{-3}$ vessel volume $h^{-1}$ have been used. Residual DO levels in the digester should not be lower than $1\,mg\,L^{-1}$.

Aerobic sludge digesters in industrial wastewater treatment plants can be designed as cyclic reactors. This can be a convenient mode of operation because excess sludge is unlikely to be wasted on a continuous basis. Furthermore the cyclic operation allows the digester to stop aeration and settle the solids therein. Fine air diffusers fitted with flexible membranes have worked well in such digesters. These diffusers would be fitted on a closed air piping grid placed on the bottom of the digester vessel. As in the SBRs, such closed air piping loops allow for quick pressure equilibration to be achieved when aeration is restarted following a phase without aeration. Supernatant can then be skimmed from top and discharged back to the headworks for further treatment while thickened sludge can drawn from the bottom of the vessel before the digester is charged with the next batch of waste activated sludge.

## 6.4. Sludge Conditioning

In the smaller IWTPs, sludge may be dewatered without prior conditioning. The argument for this is again the relatively small quantities of sludge generated. In terms of costs and simplicity in operation, it may be preferred to have a larger dewatering device to cope with the less thickened sludge (and accepting this less than optimal operation of the dewatering device).

Where conditioning is practiced, inorganic chemicals such as iron salts and lime have been used. The use of such chemicals does increase the quantity of dewatered sludge requiring final disposal. Nevertheless some use may be necessary especially for sludges which have not been stabilized. In such instances, lime has been added to reduce odor formation and putrefaction. Dosages of iron salts used have ranged from 1–5% while lime dosages have ranged from 5–20% of dry sludge solids. In comparison to the inorganic conditioning chemicals, polyelectrolytes have been more frequently used to condition sludge at IWTPs because these are typically easier to handle than the inorganic chemicals. Polyelectrolyte dosages are typically less than 1% of dry sludge solids and the chemical is dissolved in water and applied in solutions. Appropriate dosages and proper mixing are necessary to achieve the desired sludge conditioning. In-line sludge conditioning is often practiced at IWTPs. Since this takes place just before the dewatering device, the risk of prolonged mixing and hence deterioration in filterability is avoided.

## 6.5. Sludge Dewatering

The two more frequently encountered sludge dewatering devices at the smaller capacity IWTPs are the sludge drying beds and filter press. Where space is available and the sludge quantity is relatively small, the sludge drying bed has worked well. The filter press is selected when there are space constraints. Sludge drying beds are typically built in pairs so that while one is dewatering the sludge, the other can be filled (Fig. 6.5.1).

The beds comprise a layer of coarse sand laid on a layer of gravel. The latter is then laid over drains or perforated pipes resting on the concrete floor and these serve as the underdrains which collect the filtrate draining through the bed (Fig. 6.5.2). Sludge is placed on the sand layer in approximately 20 cm layers until the design loading is reached. These beds are held within low water-tight concrete walls. The loaded bed would then be left for 10–20 days to dewater the sludge. Sludge loading rates of 200–600 kg dry solids $m^2$ bed area $year^{-1}$ have been used in the tropics and the resulting sludge cake can have 20–40% solids

*Sludge Management* 103

Fig. 6.5.1. A pair of sludge drying beds with one (RHS) in drying phase while the other (LHS) is in filling phase. Splash plates need to be placed under the feed pipes to reduce disruption of the sand layer beneath.

Fig. 6.5.2. A pair of newly constructed sludge drying beds showing the gravel bed, underdrain, and sludge application pipes above.

Fig. 6.5.3. Large drying beds fitted with running tracks for mechanical cake removal. These beds serve an industrial scale piggery farm.

content. With small beds, the sludge cake is removed manually and bagged in preparation for disposal. With large beds, the sludge cake may be removed using a small bulldozer running on tracks laid on the beds (Fig. 6.5.3).

Filter presses are also called plate and frame presses. Each press comprises a number of plates mounted together to form hollow chambers. A filter cloth is mounted on each plate. This cloth serves to retain the solids while allowing the filtrate to pass through as the press is progressively filled with sludge. While the press may be filled in about 30 mins, pressure can be maintained for hours so as to force more filtrate through the cloth (Fig. 6.5.4). A variation of the fixed volume filter press is the variable volume recessed plate pressure filter. The diaphragm placed behind the filter cloth utilizes air or water pressure to squeeze the sludge and hence forcing the filtrate through the cloth. The sludge cakes produced by filter presses are dryer than that produced by the drying beds and 40–60% solids have been noted. Typically in the factory environment, the press is operated during the day shift only. One or two cycles of operations would take place within this shift.

The dewatered sludge is bagged and disposed off at a landfill periodically. Most industrial wastewater sludges are not used in composting or as soil conditioners

Fig. 6.5.4. Filter press used to dewater zinc sludge. Some of the resulting sludge cake can be seen beneath the press. The zinc has been precipitated out of the wastewater prior to anaerobic and aerobic treatment.

because of concerns over contaminants such as metals. If the dewatered sludge has not been stabilized, it has to be disposed off quickly and not stored for more than 2–3 days. This is because the organic component in the dewatered sludge may begin to putrefy and generate odors.

# CHAPTER 7

# CHEMICALS AND PHARMACEUTICALS MANUFACTURING WASTEWATER

## 7.1. Background

This is a large class of wastewaters with very varied chemical compositions and properties. The latter is due to the many different products which this group of manufacturers targets. There are some within the chemical sub-group which may not even be manufacturing the final product but are producing materials which are used by others undertaking downstream manufacturing. Pharmaceutical manufacturing using chemical synthesis have processes which bear similarities to those in the chemicals sub-group while those which depend on biosynthesis may share similarities with the fermentation industry. While the chemical and some of the pharmaceutical manufacturers may produce largely the same products continuously, many pharmaceutical manufacturers undertake campaign manufacturing as they fulfill orders for one product after another. The technical manpower required for the operation of such chemical and pharmaceutical manufacturing facilities needs to be skillful and this usually requires such facilities to be located very near or within major population centers. Consequently space constraints at site may often exist. Chemical and pharmaceutical manufacturers may also have concerns over the potential loss of confidentiality in relation to their product formulations. To protect the latter, information on formulations and specific wastewater components is comparatively rare in the published literature.

## 7.2. Chemicals and Pharmaceuticals Manufacturing Wastewater Characteristics

Given that the chemicals and pharmaceuticals manufacturing industry encompass many different types of products, different types wastewaters can be encountered within the industry. For instance organic strengths can vary from the 100s to 10 000 s mg COD $L^{-1}$. The range of wastewater pHs encountered is also very wide and wastewaters may be with and without metals. The latter could have

Table 7.2.1. An example of a pharmaceutical manufacturing facility generating 4 streams of wastewater with different properties.

| Stream No. | COD, mg L$^{-1}$ | Number of components |
|---|---|---|
| Stream-1 | 3000 | 5 |
| Stream-2 | 6000 | 8 |
| Stream-3 | 9000 | 9 |
| Stream-4 | 10 000 | 16 |

been used as catalysts in the manufacturing processes. However, most chemical and pharmaceutical manufacturing wastewaters have relatively low SS contents. The exception to this is possibly the wastewaters arising from biosynthesis based pharmaceuticals manufacturing. Organic solvents are frequently encountered in the wastewaters and most of the latter are nutrients deficient. Some of the compounds which have been encountered in chemicals and pharmaceuticals manufacturing wastewaters include methanol, ethanol, iso-propyl alcohol, butanol, methyl-isobutyl ketone, piperidine acetate, butyl acetate, hydroxypivaldehyde, toluene, hexane, branched chain fatty acids, and ethanoic acid.

An example of a pharmaceuticals manufacturing wastewater from a facility with a single stream of wastewater had an average COD of 25 000 mg L$^{-1}$, BOD$_5$ 10 000 mg L$^{-1}$, TOC 9000 mg L$^{-1}$, SS 10 mg L$^{-1}$ and pH 5. The difficulty with pharmaceutical wastewater is that it may comprise several streams with very different properties and if these are not generated continuously, blending the streams to produce a stable combined stream can be difficult. An example of this phenomenon is provided in Table 7.2.1 where a facility generated 4 streams with different properties. The fact that the 4 streams are different can be seen from the differing COD strengths of the streams. Even more indicative of the differences between the 4 streams is the number of organic components in each of the streams. These ranged from 5 in Stream-1 to 16 components in Stream-4.

The differences in properties among the 4 streams need to be noted because it is unlikely a wastewater treatment facility with four separate treatment trains — one for each wastewater stream — would be constructed.

## 7.3. Chemicals and Pharmaceutical Manufacturing Wastewater Treatment

The specific treatment strategy adopted for a particular chemical or pharmaceutical manufacturing wastewater is very dependent on the nature of the manufacturing processes used. Typically the treatment strategy includes equalization and biological treatment. A treatment train which was used to treat a wastewater with methanol as the main organic component included the following. (a) receiving sump (which also served as the equalization tank), (b) pH control, (c) nutrients supplementation, (d) twin reactor SBR, (e) excess sludge holding tank, and (f) sludge dewatering by filter press. A similar approach was used to treat wastewater from a personal care products manufacturing facility (Fig. 7.3.1), the differences being the biological treatment unit which was a twin-tank SBR in the previous example is a single tank in Fig. 7.3.1, a DAF for removal of O&G had been inserted between Stages (b) and (c), and the sludge press in (f) had been substituted with sludge drying beds. The single tank SBR in the latter was possible because the manufacturing facility operated on a single shift. This meant that the SBR vessel was receiving wastewater when the manufacturing facility was in operation but transited to the treatment phase when the manufacturing facility had shut down for the day. If the wastewater contained metals like zinc, then metal removal by chemical precipitation would precede biological treatment. Macro-nutrients supplementation is usually necessary and in addition to these, there have also been occasions when some combination of micro-nutrients has to be supplemented as well.

A single equalization vessel may not be adequate at some facilities. This is particularly so for facilities like the one shown in Table 7.2.1 where the facility

Fig. 7.3.1. Single reactor aerobic SBR for personal care products manufacturing wastewater treatment.

generated a stream for a period of time, then stopped generating that stream, and generated another for the next period of time. This is frequently associated with campaign manufacturing activities. Since treatment process instability would likely occur if the wastewater treatment facility was fed with a very different next stream after one has ended, the streams of wastewater were held separately in their respective holding tanks when generated. These segregated streams were then bled into a mixing tank at predetermined rates to generate a much more consistent combined wastewater stream which would be fed into the next unit process. Each holding tank in the system had sufficient capacity to hold the wastewater generated by each manufacturing campaign.

## 7.4. Chemical and Pharmaceutical Manufacturing Wastewater Treatment Issues

Biological process inhibition is among the more frequently encountered difficulties associated with the treatment of chemicals and pharmaceutical manufacturing wastewaters. Such inhibition may be caused by the presence of inhibitory organics and metals. It should, however, be noted that inhibition is not just a function of the organic or inorganic substances present but is dependant on its concentration and other environmental factors. Two of the latter are the pH and TDS concentration of the wastewater. Both high and low pHs can be encountered and either one can contribute to inhibition. High TDS concentration can cause dehydration of the microbial cells and interfere with settling of the biomass in the reactor. TDS concentrations of about 3500 mg $L^{-1}$ may already begin to cause process difficulties.

A key approach towards addressing the potential for inhibition caused by pollutant components present in a wastewater is the segregation of wastewater streams and holding these in separate tanks. The contents of these tanks are then withdrawn and carefully blended to produce as stable (in terms of composition) a combined stream as possible and one which has a lower inhibition potential because of the dilution of the stream containing the inhibitory substance.

Failure to do this can easily result in process instability and a consequence of the latter can be bulking sludge. A normal biomass has some filamentous growth and these form the structure upon which biomass may flocculate (Fig. 7.4.1). The biomass may also show the presence of higher animals which are grazers. A well flocculated and compacted biomass can have a sludge volume index (SVI) of 100 or lower. A consequence of inhibition can be the rapid growth of filamentous micro-organisms (Fig. 7.4.2) resulting in the phenomenon known as bulking sludge. The SVI can then exceed 150. A further sign of inhibition can be the disappearance of higher animals. Bulking sludge causes difficulties in the

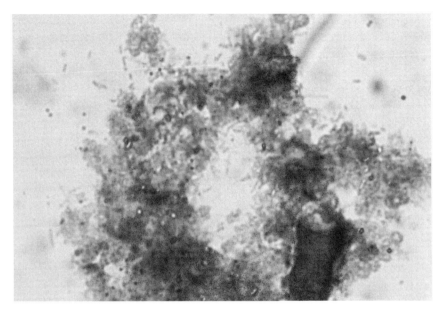

Fig. 7.4.1. Biomass from an activated sludge process operating normally. A higher animal is indicated on the lower RHS of the picture.

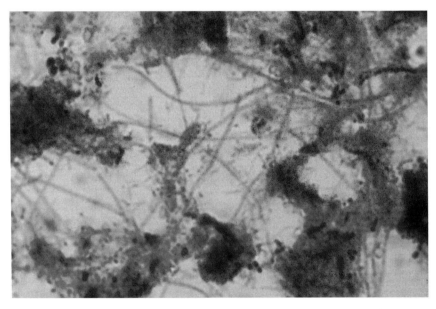

Fig. 7.4.2. Inhibited activated sludge biomass with filamentous growth. This condition may be indicative of bulking sludge.

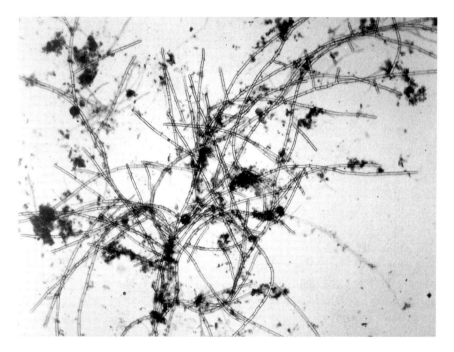

Fig. 7.4.3. Filamentous microbes in a severely bulking sludge.

liquid-solids separation stage because of the biomass's poor compaction. The latter may then result in biomass overflowing the clarifier and their loss from the system. Treated effluent SS would likely exceed the discharge limits. Difficulties can also be expected when dewatering bulking sludge. A wetter than usual cake would likely occur.

Compounds in chemicals and pharmaceuticals manufacturing wastewater need not always be difficult to degrade to cause bulking sludge difficulties. Wastewaters with easily degradable organics providing the main component of organic strength can also be difficult to treat. Examples of such organics include ethanoic acid and methanol. Such compounds can easily lead to imbalances between the carbon substrate and nutrients. Sludge bulking can then develop and SVIs can reach 200 very quickly upon treating such wastewaters (Fig. 7.4.3) and chlorination may then have to be initiated to bring the filamentous microbes under control. The SBR has been used to effectively control filamentous growth when treating such wastewaters and hence reduce the necessity for chlorination by manipulating the ratio between the unaerated portion of FILL and the aerated portion.

# CHAPTER 8

# PIGGERY WASTEWATER

## 8.1. Background

While there are large pig farming operations in Asia with tens of thousands of standing pig population (spp) per farm, these are, comparatively, few in numbers (Fig. 8.1.1). Many pig farms are relatively small operations with 10 to 1500 spp. These small farms are typically family-operated and located near population centers. In the last decade or so, efforts have been made to cluster these small farms and organize them into collectives with over 10 000 spp per collective so that their wastewaters can be handled by centralized facilities but much remains to be done. Piggery wastewater, rather than piggery wastes, is the issue because pig pens in Asia are typically not cleaned by scrapping which would have resulted in a relatively dry waste. Instead the droppings are removed from the pens by washing and flushing the latter out of the pens with water. Consequently the wastes from pig farms are liquid and should be more correctly referred to as wastewater. This wastewater is very strong, both in terms of organic strength and suspended solids content. Flushing also assists in conveying the wastes to the treatment facilities by way of drains located along the sides of the pig pens (Fig. 8.1.2). While the large farms would have a combination of sows and boars, baby pigs, weaned piglets, and growing pigs, many small farms focus on producing animals of a particular size to meet the demands of the local market.

## 8.2. Piggery Wastewater Characteristics

The amount and composition of piggery wastewater depends on the numbers of animals on the farm, their weight and age, and feed composition. Materials such as vitamins, antibiotics, and growth promoters are used in the animal feed formulations and these can appear in the wastewater subsequently. As an example, copper was used as a growth promoter in the past and copper concentrations as high as 300 mg $L^{-1}$ have been found in piggery wastewaters then. Before the flushing and

Fig. 8.1.1. Modern pig farm. Industrial-scale farms keep their pigs in pig houses such as these which are automatically flushed to clean at predetermined intervals during the day.

washing, piggery waste is composed primarily of animal feces and urine. The volume of washwater used can vary substantially from farm to farm depending on cleaning practices. For concrete-lined animal pens, this can vary from 20–45 L washwater $spp^{-1} d^{-1}$. Average values of wastewater parameters describing piggery wastewater are shown in Table 8.2.1.

The necessity for knowing the age and weight profile of the animals on a particular farm cannot be over emphasized. This is because the size of the animals can substantially impact on the amount and composition of wastes generated. To illustrate this, Table 8.2.2 provides information on animal age and weight and relates these to the amount of feces and urine produced. Young animals (<8 weeks) produce almost twice as much urine as feces while older animals (20–23 weeks) produce about equal amounts of feces and urine. This can have a substantial impact on the TKN concentration in a piggery wastewater.

From Table 8.2.1, the key characteristics of piggery wastewater are its very high SS and organic content (as indicated by its $BOD_5$ and COD values). In addition to this, ammonia is an issue given the high TKN value and the large component of urine in the wastes. This is particularly so if a farm specialized in raising very young animals for the market. In places where receiving waters may be threatened by excessive nutrients, the nitrogen and phosphorous in the wastewater would

Fig. 8.1.2. Pig pen at a small farm. These pens are usually manually flushed by the farmer. The drain conveying the wastewater to the treatment plant may be seen on the LHS of the picture. This drain also serves to drain rainwater running off the roof which is undesirable as it unnecessarily increases the hydraulic load on the treatment plant during rain storms.

Table 8.2.1. Average characteristics of piggery wastewater.

| Parameters | Average values, mg $L^{-1}$ |
|---|---|
| $BOD_5$ | 5000 |
| COD | 20 000 |
| SS | 20 000 |
| TKN | 900 |
| $PO_4$ | 200 |

be an issue. Piggery wastewater being of animal origin can be expected to have organic components which are easily biodegradable and this is evidenced by the ease with which degradation begins even in a collection sump at the end of the collection drains (Fig. 8.2.1). A collection or equalization sump is needed because wastewater would not be generated on a continuous basis. The pens are likely to be washed and flushed twice a day during working hours. This means there would be two sharp spikes in wastewater flows and relatively very low flows in between.

Table 8.2.2. Piggery waste quantity and composition with respect to animal age and weight.

| Animal age (weeks) | Animal wt (kg) | Feces (g d$^{-1}$ kg$^{-1}$ animal) | Urine (g d$^{-1}$ kg$^{-1}$ animal) | Total wastes wt (g d$^{-1}$ kg$^{-1}$ animal) | Total wastes volume (L d$^{-1}$ kg$^{-1}$ animal) |
|---|---|---|---|---|---|
| 8< | 18< | 27 | 58 | 85 | 46 |
| 8–12< | 18–36< | 43 | 48 | 91 | 50 |
| 12–16< | 36–54< | 54 | 61 | 115 | 63 |
| 16–20< | 54–72< | 46 | 58 | 104 | 57 |
| 20–23 | 72–90 | 47 | 51 | 98 | 53 |

Fig. 8.2.1. Piggery wastewater collection sump. The "frothy" appearance on the surface of the wastewater was caused by the release of gases arising from fermentation. A large component of the organic pollutants in piggery wastewater is easily biodegradable.

## 8.3. Piggery Wastewater Treatment

Where small farms handle their wastewater treatment on an individual basis, it is unlikely that much thought can be given to resource recovery since economy of scale is unlikely to occur. The treatment train would typically include the removal of coarse particles, equalization, organic strength reduction, (probably with an anaerobic process), removal of organics to discharge limits and nutrients removal where necessary (probably with an activated sludge process variant), and sludge treatment.

The coarse screens located at the beginning of the treatment train serve to protect mechanical equipment downstream. These screens are important because large pieces of various materials, including baby pigs, have been known to enter the drains. The coarse screens may be followed by fine screens and these serve to partially replace primary clarifiers in terms of solids removal. Operating primary clarifiers can be difficult because of the wastewater's biodegradability. During periods of low flow, hydraulic retention times consequently become longer than design values and septic conditions can develop with the release of gas.

When the treatment train includes continuous flow unit processes, there is a need to equalize (or hold) the wastewater so that wastewater from the two discharge episodes can be spread over the day. Holding piggery wastewater is not, however, a desirable option because of its biodegradability. Because of this, holding tanks are typically designed small or eliminated. At places where holding tanks are not included, then the receiving continuous flow unit process should have sufficient buffering capacity. Due to this latter requirement, many treatment trains include lagoons. Anaerobic lagoons have been used to reduce organic strength. These are constructed up to 4 m water depth with loading rates of 0.08–0.16 kg VS m$^{-3}$ lagoon volume or 5.0–6.5 m$^3$ 100 kg$^{-1}$ spp. The operating pH is typically 6.8–7.4 and at least intermittent mixing (can be by recirculating lagoon liquor) is desirable. Anaerobic lagoon effluent BOD$_5$ can be expected to be 1000–2000 mg L$^{-1}$ while N and P can be about 800 mg L$^{-1}$ and 80 mg L$^{-1}$ respectively. The resulting BOD:N:P ratio of 25:10:1 means the lagoon effluent is amenable to further aerobic treatment. Odor is a problem with anaerobic lagoons and this is particularly so if the lagoons have not been desludged regularly. Sludge accumulation can be as high as 0.15 m$^3$ spp$^{-1}$ year$^{-1}$.

Figure 8.3.1 shows a set of small anaerobic lagoons (or ponds). While not evident from the picture, these have malfunctioned (ie. effluent not meeting design values) as a result of inadequate desludging.

The anaerobic lagoon may be followed with the facultative lagoon. This is 2–3 m deep with loading rates of up to 300 kg BOD$_5$ ha$^{-1}$ d$^{-1}$. Hydraulic retention times are typically 10–15 d and BOD removal can be 70% or better. Sludge accumulation is also an issue although (at 0.05–0.08 m$^3$ spp$^{-1}$ year$^{-1}$) not as serious as in anaerobic lagoons.

An aerobic process, such as the aerated lagoon, then follows to remove the organics to discharge limits and to nitrify the wastewater. Aerated lagoons are

Fig. 8.3.1. Anaerobic ponds (which have been staged) for piggery wastewater treatment.

constructed with depths of 3–5 m and loading rates can vary widely — from 0.6–6 m$^3$ lagoon volume 100 kg$^{-1}$ spp. BOD removal and TKN conversion of 90% and 80% respectively can be expected. Apart from the aerated lagoon, other activated sludge variants such as the oxidation ditch and conventional activated sludge process have also been used. These are operated with sludge return flows of 25–50% of the influent flow. Figure 8.3.2 shows the oxidation ditch variant where a pair of ditches has been installed. A third variant is the aerobic sequencing batch reactor (SBR) (Fig. 8.3.3). This variant, being a cyclic reactor, is well suited to be matched with an anaerobic cyclic reactor, the anaerobic SBR which would perform the initial organic strength reduction.

With the anaerobic SBR, the treatment train need not have the holding tank since its operating sequence can be designed to match the wastewater discharge pattern. With a pair of anaerobic SBR, each can be designed to have a FILL phase which matches one of the discharge episodes. Loading rates range from 3–4 kg VSS m$^{-3}$ d$^{-1}$ or 6–7 kg COD m$^{-3}$ d$^{-1}$.

Mixing is important to improve contact between the biomass and substrate and this is achieved by a combination of the biogas generated and by recirculating reactor liquor from the top third of the reactor. Recirculation is also important since it helps to release the gas bubbles from the biomass particles. Failing this,

Fig. 8.3.2. Oxidation ditch system for (anaerobically) pretreated piggery wastewater. This is a pair of oxidation ditches fitted with rotor brushes for aeration.

Fig. 8.3.3. A cyclic piggery wastewater biotreatment system. The anaerobic SBR (LHS — With the red-mud membrane cover) is followed by the aerobic SBR (RHS — Without the cover).

biomass particles may be buoyed up to the surface of the reactor's liquor forming a scum layer. This scum layer would compromise the reactor's designed hydraulic capacity and can eventually result in biomass being washed out of the reactor during the DECANT phase. The anaerobic SBR can be operated with flocculant biomass and this forms an expanded sludge blanket when recirculation occurs. The reactor side water depth is 4–5 m and the cover of the reactor can be of a collapsible material (such as a synthetic material membrane). The collapsible membrane cover serves to collect the biogas generated and by collapsing during DECANT helps to prevent ingress of air (Fig. 8.3.4). The feed pipes also serve as the recirculation flow pipes and these pipes ensure the flow is spread as evenly as possible across the sludge blanket (Fig. 8.3.5). Biogas from the anaerobic SBR has been collected for heating purposes during animal feed preparation at the smaller farms.

The anaerobic SBR can be followed with the aerobic SBR. This, like the other aerobic treatment options is designed with (equivalent) F/M ratios ranging from 0.25–0.50 BOD $MLSS^{-1} d^{-1}$. Alternatively the sizes of these aerobic systems may be estimated by considering loadings of 30–80 kg BOD 100 $m^{-3}$ reactor volume $d^{-1}$. The sludge residence time is typically at least 5d and certainly not less than 3d. Membrane diffusers (Fig. 8.3.6) have been successfully used in such plants with few diffuser clogging problems. BOD removal by any of the aerobic options can be expected to be at least 95%.

Often waste sludge (eg. primarily from the aerobic process) is dewatered using sludge drying beds. The larger sludge drying beds (Fig. 6.5.3) may be equipped with tracks which allow a small excavator like a "Bobcat" to enter the bed and shovel the sludge cake out of it. The smaller beds on small farms are typically managed manually.

## 8.4. Piggery Wastewater Treatment Issues

Foaming can be an issue during plant start-up of the aerobic component. It seems particularly prevalent when a relatively small amount of seeding material had been used and wastewater loads on reactors are increased too quickly. When very severe foaming occurs (Fig. 8.4.1), application of a silicone based anti-foam agent has been found to be helpful. The loss of biomass during foaming episodes inevitably result in uncontrolled biomass loss and hence would further delay completion of process start-up. Foaming has not been noted to be an issue after a plant has passed the start-up phase.

Highly colored treated effluent can also be an issue and this is an issue which has often been associated with agricultural and agro-industrial wastewaters. The

Fig. 8.3.4. Anaerobic SBR collapsible membrane cover detail. The set of pipes on the LHS are for the feed and recirculation system.

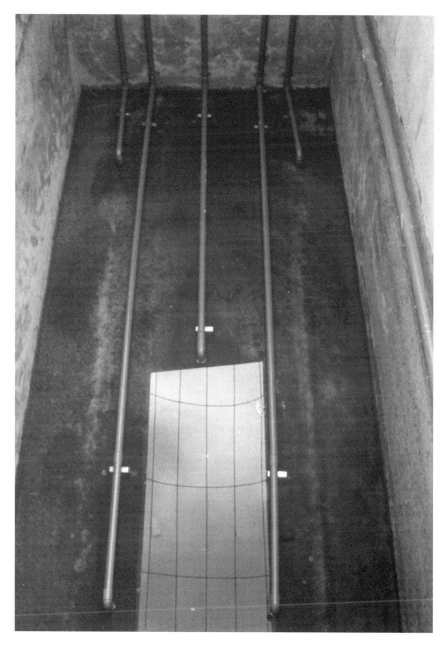

Fig. 8.3.5. Anaerobic SBR internal details showing the feed/recirculation piping.

Fig. 8.3.6. Diffuser grid in an aerobic SBR (drawn down for maintenance). Membrane diffusers have been selected to reduce the incidence of diffuser clogging.

dark brown coloration (Fig. 8.4.2) has been linked to the feed formulation used and the SRT of the aerobic process. Modifying the feed formulation (eg. reducing the amount of molasses used) and shortening the SRT has proved helpful in reducing the color problem (Fig. 8.4.3). Highly colored treated effluents have been associated with systems operated with lower and longer than expected loadings and SRTs respectively, and although BODs can be very low, CODs can exceed discharge limits

Final disposal of the dewatered sludge has also been an issue. There is some scope for its use as a soil conditioner and this has occurred at locations where other

Fig. 8.4.1. Severe foaming during start-up of an aerobic SBR unit treating piggery wastewater.

Fig. 8.4.2. Strongly colored treated piggery wastewater. This would have a low BOD but a COD which exceeded the discharge limit.

Fig. 8.4.3. Relatively color-free treated piggery wastewater. This followed reformulation of the animal feed.

crop growing farms adjoin the pig farms. An example was an areca palm (*areca catechu*) plantation receiving treated effluent for irrigation and dewatered sludge for soil conditioning from an adjoining pig farm but the quantities involved had been small. The issue can be one of distance between the two farms. In a number of Asian communities, there are also possibly religious sensitivities concerning the use of pig farm derived treated effluent and dewatered sludge for application at farms raising food crops.

# CHAPTER 9

# SLAUGHTERHOUSE WASTEWATER

## 9.1. Background

Slaughterhouses in Asia largely slaughter poultry (with chickens exceeding ducks) and pigs. Cattle, sheep, and goats are slaughtered in smaller numbers. Many of these slaughterhouses are small and serve the communities in their vicinities. These slaughterhouses receive their animals late in the evening, slaughter them in the very early morning (eg. 0000–0400 h), and by early morning (eg. 0400–0600 h) the animal carcasses would have been delivered to the market. The carcasses may even be delivered to the market warm rather than chilled if distances are short. Since these slaughterhouses receive their live animals daily, they do not hold their live animals for long and are unlikely to have very large holding pens (Fig. 9.1.1). The number of animals (eg. pigs) slaughtered daily may be as few as in the tens. Such slaughterhouses may have no activities during the daytime. Large slaughterhouses may, however, operate continuously throughout the day. Consequently they require substantial animal holding facilities. This means such slaughterhouses would not only have a wastewater stream from the slaughtering activities but also a wastewater stream which bears resemblance to farm wastewater. Poultry slaughterhouses often do not have holding pens. The poultry arrive at the slaughterhouse in cages (Fig. 9.1.2) and since the interval between arrival and entering the process line is usually relatively short, the birds remain in these cages before being slaughtered.

The activities in the pig slaughter process can include (a) drawing the pigs from their holding pens, (b) stunning the animals, (c) throat cutting and blood letting, (d) hot water scalding, (e) dehairing, (f) eviscerating, (g) bowel washing, (h) meat inspection for parasites, and (i) transportation to market. Among these activities — throat cutting and blood letting, hot water scalding, dehairing, eviscerating, and bowel washing have potential for generating waste and wastewater streams. Throat cutting and blood letting can result in a very strong wastewater stream if blood is not adequately collected. This is a stream which has a high organic and nitrogen content. The blood may be recovered as part of the feed for a

Fig. 9.1.1. Animal (pigs) holding pens at a small slaughterhouse.

Fig. 9.1.2. Bird cages at a poultry slaughterhouse.

Fig. 9.1.3. Viscera cleaning following pig slaughter. Washings from cleaning the work surfaces flow into the drain on the RHS.

rendering plant or in Asia it may be prepared for the market since it is consumed. Bowel washing is frequently encountered at slaughterhouses in Asia because there is again a market for the viscera. This is usually performed manually (Figs. 9.1.3). Bowel washing frequently uses a great deal of water and results in a very strong organic wastewater stream with substantial quantities of particulates. Water is, of course, used at regular intervals to wash the process lines and work floor to ensure good hygiene. All these streams are combined and become the slaughterhouse's wastewater. A poultry slaughter line largely follows the same unit process flow scheme as the pig slaughter process but with variations such as the dehairing step being replaced with defeathering.

## 9.2. Slaughterhouse Wastewater Characteristics

The amount and composition of slaughterhouse wastewater obviously depends on the number and type of animals processed. Housekeeping practices can significantly impact on the volume of wastewater generated. At the smaller slaughterhouses it may be necessary to regularly flush the work surfaces and floors with water to clear materials such as blood which may have dripped from the

Fig. 9.2.1. Animal blood on the floor at a slaughterhouse. The floor needed to be cleaned regularly during the hours of slaughterhouse operation to ensure adequate hygiene and to keep the floor from becoming slippery.

slaughtered animals (Fig. 9.2.1). Table 2.1.1 provides some values of wastewater characteristics for poultry slaughterhouses while Table 2.3.4 provides wastewater generation estimates for poultry and pig slaughterhouses. A number of features may be observed from the tables and these include the low BOD:COD ratio — suggesting an easily biodegradable wastewater, relatively low BOD:N ratio — suggesting a need for nitrification and possibly even nitrogen removal, and relatively high SS. The TKN values are very dependent on how blood is handled at a slaughterhouse. As far as is practicable, blood should be carefully collected and disposed off separately from the wastewater. This is because blood itself can have BOD values of about $100\,000\,\text{mg}\,\text{L}^{-1}$. Similarly SS values are very dependent on bowel washing activities. If viscera are not recovered for the market, then SS in slaughterhouse wastewater can be low. If bowel washing occurs, then this stream should be pretreated to reduce SS content before allowing it to join the main stream of wastewater flowing to the wastewater treatment facility. Do note that the wastewater values shown in the tables are for wastewaters which have undergone coarse screening and hair or feathers would have already been removed.

As pointed in Sec. 9.1, smaller slaughterhouses serving local markets do not operate continuously throughout the day. Typically they operate during the night and there can be no activity during the day. This means wastewater generation largely occurs during the night.

## 9.3. Slaughterhouse Wastewater Treatment

Coarse screens need to be located at the beginning of the treatment train. This is to ensure gross particles like hair, feathers, and discarded parts of the slaughtered animals do not enter the treatment system and damage the mechanical equipment. It should be noted that there may be commercial value in the hair and feathers and these are then recovered at the slaughterhouse. The bulk of the hair and feathers may not appear in the wastewater (Fig. 9.3.1). Often the coarse screens are followed by macerating pumps which would lift the wastewater from the collection sump onto the next unit process. The macerating pumps provide additional protection (for the equipment located downstream) from gross particles by reducing the size of these. Fine screens ($\sim 1$ mm aperture) may be provided after the

Fig. 9.3.1. Resource recovery from slaughterhouse wastes — feathers. The feathers are collected in the bamboo baskets in the foreground of the picture. Typically the recovered feathers are removed from the slaughterhouse daily.

pumps. These help reduce the organic and solids load on downstream unit processes allowing them to be sized smaller. In places where they are provided, these fine screens may be either of the mechanical rotary or self-cleaning curved type. Both types have been used successfully. The wastewater can be expected to contain some O&G and when this is mixed with the blood and fine SS, a scum result (Fig. 9.3.2). This mixture of material can easily result in odorous conditions if housekeeping at the wastewater treatment plant is not well practiced. The O&G and particularly the blood can very substantially increase the aeration needs of the aerobic process. At the larger slaughterhouses, coagulation-flocculation followed by dissolved air flotation is often used to remove these two wastewater components. Clarifiers are rarely used because the longer residence times in these can result in septic and consequently odorous conditions developing. Aluminum or iron salt coagulant have been successfully used to assist SS removal. Adequately designed DAFs can be expected to remove 60% or more of the nitrogen, O&G, and SS. While DAFs have been found effective when operated appropriately, many have also been found to perform below design expectations. A key reason for this lower than expected performance, is the inherent instability in the coagulation/flocculation-DAF processes if this is not operated continuously.

Fig. 9.3.2. Wastewater collection sump at a poultry slaughterhouse. The scum comprised feathers, O&G, and blood.

Since many slaughterhouses do not operate continuously throughout the day, their wastewater treatment plants would not receive wastewater continuously and the coagulation/flocculation-DAF processes may be shut down when a slaughterhouse is not operating and restarted when wastewater is next generated. Possible solutions to this problem would be to have a small wastewater holding capacity to serve as a buffer and to allow effluent from the coagulation/flocculation-DAF process, following a "restart", to be recirculated to the holding tank until DAF effluent quality has stabilized. Then the DAF effluent is allowed to continue into the biological process. Alternately a larger holding tank is provided to allow the wastewater treatment plant to be operated continuously throughout the day. While both approaches have been found to work, the latter approach would obviously allow the unit processes downstream of the holding tank to be sized smaller since the flow would now be spread out over 24 h. The larger holding tank would, however, need to be well mixed and its contents prevented from turning septic and odorous. Mixing with diffused air is the preferred option.

Aerobic biological treatment options which have been successfully used include the activated sludge process, oxidation ditch, and sequencing batch reactor. Notwithstanding the use of DAF to pretreat the wastewater, the latter's nitrogen content must still be taken into consideration with the wastewater's BOD content when estimating aeration requirements. This is because typically at least nitrification is a treatment requirement and nitrification exerts a high oxygen demand. Where nitrogen removal is a requirement, then nitrification is followed with denitrification. Of the three biological treatment configurations identified, two — the oxidation ditch and sequencing batch reactor — can be designed to allow this to happen quite readily. In places where there are space constraints, the SBR instead of the oxidation ditch has been used effectively.

At locations where space is not as constrained, lagoons have also been successfully used. Such treatment trains also begin with screens — coarse and possibly fine screens. DAFs may not be used then. Anaerobic lagoons are likely to follow and these can be carried out in two stages with the first stage loaded at 0.7 kg $BOD_5$ m$^{-3}$ lagoon volume. The second stage anaerobic lagoon can be loaded at 0.2 kg $BOD_5$ m$^{-3}$ lagoon volume. The BOD removal by the first and second lagoons would be about 65% and 60% respectively. Total BOD removal by the two lagoons can be expected to be about 85% of wastewater BOD. The anaerobic lagoons would be followed by an aerated lagoon and this is loaded at about 0.07 kg $BOD_5$ m$^{-3}$ lagoon volume. The HRT would be about 2–3 days. The aerated lagoon can be expected to remove about 80% of the BOD entering it. Surface aerators are typically used to aerate the lagoon's mixed liquor. These surface aerators can either be float mounted or pier mounted (Fig. 9.3.3).

Fig. 9.3.3. An aerated lagoon at a slaughterhouse. The three sets of piers are for mounting surface aerators. The animal holding pens are on the left hand side of the picture.

## 9.4. Slaughterhouse Wastewater Treatment Issues

Many slaughterhouse wastewater treatment plant failures can be traced to process instabilities resulting from variations in the wastewater stream. These variations result primarily from the manner in which the slaughterhouses are operated. There may be no slaughtering activity during the day or on certain days and this would mean no or little wastewater flow during such periods. Unit treatment processes may have difficulty reaching "steady-state" operating conditions given such wastewater flow regimes. The problem is compounded by the very sharp increases in the number of animals slaughtered just before festivals. Such increases have been known to be two to three-fold over the usual daily average.

Asian slaughterhouses may also show a high degree of viscera recovery since there is often a local market for this. Viscera recovery necessitates viscera washing at the slaughterhouse and this result in a stronger and larger wastewater flow. This is particularly so with suspended solids and these solids can be abrasive to downstream pumps and valves if not removed. They also lead to inert material accumulation in the bioreactors.

Locating a rendering plant next to a slaughterhouse would help in reducing the pollutant load in the wastewater since much of the blood and solids would be

Fig. 9.4.1. A higher than usual amount of blood in the wastewater (red coloration). This was the result of poorer than usual recovery at the slaughterhouse.

recovered for the rendering plant. However, many slaughterhouses in Asia are not constructed with rendering plants next to them and where they exist, there has been resistance from neighbors because of the odors emanating from such rendering plants. Fortunately there is a market for blood and this is largely recovered. Nevertheless some blood still flows out of the slaughterhouse with the wastewater. Provided that the coagulation/flocculation-DAF process and pH control therein are performing to design expectations low pH is not an issue. Nevertheless, it would be prudent to anticipate downwards pH excursions in the event of poor blood recovery at the slaughterhouse or removal by the DAF resulting in increased nitrification activity in the aeration basin (Fig. 9.4.1). Vessels, and in particular steel vessels, should be carefully treated with anti-corrosion paints.

# CHAPTER 10

# PALM OIL MILL AND REFINERY WASTEWATER

## 10.1. Background

The palm oil industry has two components — milling and refining. Palm oil mills are typically located within or close to palm oil plantations. This is to ensure the harvested palm oil fruit bunches can reach the mills quickly where the palm oil may be extracted before their quality deteriorates. Since the palm oil mills are associated with the plantations, they are located away from urban centers.

The palm oil is extracted from the fresh fruit bunches (FFB) in a five stage process — steam sterilization, fruit bunch stripping, digestion, oil extraction from the mesocarp, and clarification. Wastewater generation is associated with these five stages. Oil would be a key component in these wastewater streams. Apart from these wastewater streams, oil may also enter the wastewater treatment plant because of spillages within and around the mill, and oil dripping from the fruit loading bays at the mill as the act of unloading the FFBs from trucks when these arrive at the mill inevitably subjects the fruits to some pressure and oil is then squeezed out (Fig. 10.1.1).

The crude palm oil is a mixture of glyceride esters derived primarily from fatty acids with 16 to 18 carbon atoms. Some gum is also present in the crude palm oil. Before this oil can be used by consumers it has to be refined. The latter means fractionation by physical or chemical means to produce olein, stearin, and acid oils. The physical refining process includes degumming with acid and pre-bleaching, as well as deacidification and deodorization. The last step yields fatty acid distillates, palm olein, and palm stearin. Physical refining wastewaters are made up of cooling water bleed, floor wash, equipment cleaning water, and spillages. The chemical refining process includes alkali neutralization which yields soap stock or neutralized palm oil, olein and stearin. The soap stock can undergo acidification to yield acid oil while neutralized palm oil, olein and strean can be bleached, and deodorized. Chemical refining wastewaters include streams similar to those found

Fig. 10.1.1. Palm oil FFB receiving bay. The pressure of piled up fruit bunches as these are unloaded onto the loading bay above resulted in some oil being squeezed out and this dripped onto the ground shown on this picture. This oil is collected by surface drains and at least a portion of it would make its way to the wastewater treatment plant.

in physical refining and others such as alkaline neutralization wash water, soap stock splitting effluent, and spent fractionation blowdown.

Crude palm oil is transported from the mills to the refineries by tankers. But unlike the mills, the refineries are typically located near urban centers to be close to the end-users or near the seaports to facilitate the import of crude palm oil and export of refined oil. Refineries are also often located near rivers to facilitate access to a freshwater supply which can be used for cooling purposes within the refinery. Palm oil refineries are rarely located near palm oil mills, and space constraints can often exist at palm oil refineries.

## 10.2. Palm Oil Mill and Refinery Wastewater Characteristics

In the literature palm oil mill effluent is often referred to as POME. This is a very strong wastewater in terms of organic content (Table 10.2.1) and has a thick brownish appearance. Not unexpectedly, the wastewater has high oil content. It also has a low pH and is slightly nutrients deficient (BOD:N:P at 100:3.5:0.5) if wastewater is to be treated aerobically. The POME characteristics shown in Table 10.2.1 are average values and actual values at a mill can be influenced by the quality of the fruits harvested. The wastewater is hot and this makes it more difficult to directly treat it aerobically since oxygen transfer would be less efficient.

Palm oil refinery effluent (PORE) characteristics depend on which refining method, physical or chemical, is used at a refinery. Many refineries practice both refining methods but the amount processed using each method may vary from refinery to refinery (and within the refinery as well) depending on the demand for a given grade of oil at a particular moment. Table 10.2.2 provides examples of

Table 10.2.1. Palm oil mill effluent (POME) characteristics.

| Parameters | Average values |
|---|---|
| $BOD_5$ | 23 000 mg $L^{-1}$ |
| COD | 55 000 mg $L^{-1}$ |
| TN | 650 mg $L^{-1}$ |
| TP | 120 mg $L^{-1}$ |
| Oil | 10 000 mg $L^{-1}$ |
| Volatile fatty acids | 1000 mg $L^{-1}$ |
| pH | 4–5 |
| Temperature | 45–70°C |

Table 10.2.2. Palm oil refinery effluent (PORE) characteristics.

| Parameters | Physical refining | Physical and chemical refining |
|---|---|---|
| Temperature, °C | 28–44 (35) | 42–70 (57) |
| pH | 3.8–7.0 (5.3) | 1.2–7.0 (3.0) |
| $BOD_5$, mg $L^{-1}$ | 50–1500 (530) | 1420–19 600 (4200) |
| COD, mg $L^{-1}$ | 1000–3000 (890) | 4000–33 100 (7700) |
| TS, mg $L^{-1}$ | 20–2000 (580) | 2500–45 000 (15 000) |
| SS, mg $L^{-1}$ | 20–1000 (330) | 425–2000 (1100) |
| TN, mg $L^{-1}$ | 20–1000 (50) | 0.1–2.5 (6.0) |
| TP, mg $L^{-1}$ | 1.0–600 (4.0) | 8–16.5 (12.0) |
| O&G, mg $L^{-1}$ | 25–600 (220) | 400–16 500 (3600) |

PORE characteristics for the different refining practices. PORE pH is also low but can vary over a larger range compared to POME pH. Organic content in terms of BOD, COD, and oil is high but not as high as POME. Values do, however, vary over a wider range depending on the ratio of oil produced using physical and chemical refining at a given site. PORE generation rates, especially from physical refining, are low compared to POME (2–3 m$^3$ tonne$^{-1}$ oil extracted). Physical refining can be expected to generate about 0.2 m$^3$ tonne$^{-1}$ oil processed while chemical refining would generate about 1.2 m$^3$ tonne$^{-1}$ oil processed. Often while physical refining can be encountered on its own, chemical refining would be found with physical refining.

## 10.3. Palm Oil Mill and Refinery Wastewater Treatment

Since POME is slightly nutrient deficient for aerobic treatment and due to its high organic strength, it is typically treated anaerobically first. The raw POME BOD:N:P ratio of 100:3.5:0.5 is sufficient for anaerobic treatment. This reduces POME's organic strength very substantially and result in an effluent with a BOD:N:P ratio of about 100:7:1.4. The anaerobically treated POME is amenable to aerobic treatment. Since palm oil mills are located away from urban centers, space for the treatment facility is not usually an issue. Consequently lagoons appear frequently in such treatment trains. Figure 10.3.1 shows an anaerobic

Fig. 10.3.1. Anaerobic lagoon for POME treatment. The dark brown scum layer which appears almost "crusty" is useful for excluding air from the anaerobic process occurring beneath it.

lagoon at a palm oil mill. Such lagoons are usually staged with the first of two stages providing a HRT of 60 days and the second 40 days. The effluent quality can range from 200–1000 mg $BOD_5$ $L^{-1}$. At the larger mills the lagoons, aside from being staged, may also be split into parallel units. This is to facilitate maintenance should such a requirement arise. To facilitate quick estimation of area requirement for anaerobic lagoon construction a minimum of 330 $m^2$ is allowed for each tonne FFB processed in an hour. For example 60 tonne FFB $h^{-1}$ mills have been noted to have anaerobic lagoons occupying 2–4 ha. Given the volumetric size of the lagoons, they provide sufficient buffering capacity and separate equalization capacity is not provided ahead of the lagoons. The anaerobic lagoons are, however, often preceded by oil separators. These include both the simple baffled tanks and the more sophisticated inclined plate separators. Oil removal reduces the load on the anaerobic lagoons and the amount of oil which can penetrate the treatment facility. The amount of residual oil which can be in the final effluent is often low if it is to meet the discharge limits. The anaerobic lagoon in Fig. 10.3.1 shows a scum layer. So long as this scum layer is not excessively thick it is helpful in keeping air out of the lagoon's contents. If, however, it becomes too thick then it adversely affects the effective volume of the lagoon.

When energy, and hence biogas, recovery is not an objective, the anaerobic lagoons have worked well. Even when biogas is not recovered, palm oil mills can still be net energy producers. This is because sufficient energy for milling purposes can be derived from burning the empty fruit bunches, waste mesocarp, and nut shells. Burning these waste materials would not, however, provide sufficient energy if the mill is to generate electricity which can be sent out of the mill to be used elsewhere (eg. at a palm oil refinery if one is located nearby). If such a requirement exists then the anaerobic lagoons are replaced with anaerobic digesters. These are typically mesophilic two stage digesters. Each digester is operated with a HRT of at least 7 days. The combined HRT of the anaerobic digesters is at least 15 days and given this amount of holding time a BOD removal in excess of 80% can be expected. Loading imposed on the digesters is about 4.8 kg VS $m^{-3}$ digester volume $d^{-1}$. The digesters are usually operated at temperatures of 44–52°C. This can be achieved without supplementary heating because POME is discharged hot (45–70°C) and if the digesters are lagged with insulating material some of the heat in the incoming wastewater can be retained. Given such digesters, a 60 tonne FFB $h^{-1}$ mill operating for 20 h $d^{-1}$ would require two 4200 $m^3$ digesters. There have also been attempts to operate the digesters at thermophilic temperatures but the digesters have been found more prone to instability and hence more difficult to operate satisfactorily.

Thermophilic digesters have been operated at total HRTs of about 10 days. Given the much shorter HRTs of the thermophilic and mesophilic digesters compared to the anaerobic lagoons, good contact between the anaerobic biomass and substrate is important. This necessitates adequate mixing to be provided. The viscous nature of POME makes a digester's contents difficult to mix and if mechanical mixing with stirrers had been selected then 14 kW 1000 m$^{-3}$ digester volume need be provided. Gas mixing using a combination of a central draft tube and gas lances would require only 1.8 kW 1000 m$^{-3}$ digester volume. Apart from the lower energy requirements of the gas mixing option, it has also been found to be less prone to mechanical breakdowns. Biogas yields can be as high as 0.9 L g$^{-1}$ BOD degraded and its methane content can be as high as 60%. The 60 tonne FFB h$^{-1}$ mill would have produced about 20 000 m$^3$ d$^{-1}$ biogas.

The anaerobic lagoons and digesters are frequently followed by aerated lagoons (Fig. 10.3.2). Activated sludge systems may also be used although less commonly. These aerobic systems are typically operated in the extended aeration mode with sludge residence times of 20–30 days and MLSS concentrations of about 5000 mg L$^{-1}$. Given influent BOD$_5$ of about 2000 mg L$^{-1}$, effluent BOD$_5$ values 50–100 mg L$^{-1}$ can be expected. Very little oil should reach the aerated lagoon or activated sludge plant. Failing this adequate oxygenation of the reactor's liquor can be adversely affected. The reactor is also expected to be operating at temperatures of about 30°C. This means it is necessary to ensure the wastewater

Fig. 10.3.2. Aerated lagoon treating POME. Typically such rectangular-shaped lagoons are designed to be aerated with at least 2 surface aerators for better mixing of the lagoon's liquor.

entering the reactor is not excessively hot. The latter is, however, an unlikely event if the anaerobic pretreatment stage is of the mesophilic type.

As with POME, PORE treatment begins with oil removal. This may begin in the drains in the refinery and those leading to the wastewater treatment plant. There would also be at least a baffled tank oil trap just before the wastewater enters the treatment plant. Unlike POME treatment plants, PORE treatment plants are often equipped with DAFs for additional oil removal before biological treatment. This is particularly so if the refinery practiced chemical refining. The DAF is typically coagulant assisted and alum is commonly used in Southeast Asia.

At refineries without chemical refining, the treatment plant may not include a DAF. The incoming wastewater, after the oil trap, would have its pH corrected (Fig. 10.3.3). Physical refining produces less wastewater and one which is also

Fig. 10.3.3. pH correction station at a PORE treatment plant. This is a single stage pH correction station with sodium hydroxide.

weaker and a single-stage pH correction station has been found adequate. Sodium hydroxide has been used as the alkali, primarily for ease of handling.

Aerobic treatment follows the oil trap and pH correction station. Variants of the activated sludge process have been commonly used and these include the conventional activated sludge process, oxidation ditch, and the aerobic SBR. The aerobic SBR has been a common aerobic treatment process at refineries which do not practice chemical refining. In places where SBRs have been used these have typically been twin-tank systems (Fig. 10.3.4) operated at 6–8 h $cycle^{-1}$. MLSS concentrations used have ranged from 3000–6000 mg $L^{-1}$. Loading on the reactors have ranged from 0.1–0.3 kg $BOD_5$ $kg^{-1}$ MLSS $d^{-1}$. While BOD removal appeared little affected by MLSS concentrations within the range indicated, COD removal had been noted to be better at higher MLSS concentrations.

Fig. 10.3.4. Twin-tank aerobic SBR for PORE treatment. The third vessel on the LHS is a treated wastewater storage tank. Treated effluent quality from the SBRs had been found adequate at this refinery for use in its cooling towers. The rectangular structure in the foreground on the left hand side is a pair of sludge drying beds. Treatment of physical refining wastewater does not generate large quantities of sludges. This is unlike chemical refining wastewater which requires coagulant assisted dissolved air flotation before biological treatment. The use of coagulants results in much larger quantities of sludge requiring disposal.

## 10.4. Palm Oil Mill and Refinery Wastewater Treatment Issues

Although POME has been successfully treated, treatment plants have also faced difficulties. A common problem is the inadequate maintenance of the lagoons. Bund erosion and sludge accumulation can reduce the effective volumetric capacity of the lagoon. This reduces the HRT and hence the organics removal effectiveness. Figure 10.3.1 shows a lagoon with eroded bunds. With the bunds eroded, surface runoff can find its way into the lagoon when it rains and this would again reduce HRTs. In POME treatment, anaerobic lagoons are rarely desludged. They are instead used till sludge accumulation becomes an issue. A new lagoon is then constructed while the existing one is decommissioned and backfilled with the sludge still in it.

In places where anaerobic lagoons have been constructed in soils with magnesium sulphate, the ammonium phosphate in POME may react with the latter and form $NH_4MgPO_4 \cdot 6H_2O$ crystals. These crystals may accumulate in the pump casings and pipes reducing their capacities and causing wear in the mechanical equipment.

Although much biogas may be recovered from the anaerobic digestion of POME, the biogas is corrosive and can damage boilers and generator sets. This corrosion problem can be overcome by selecting an appropriate lubricating oil which can neutralize the corrosive components in the gas.

While the aerated lagoons which follow the anaerobic lagoons or digesters are usually robust, their performance has been known to decline sharply should excessive amounts of oil penetrate the anaerobic stage and enter the aerobic stage. Such oil excursions may occur because of spillages at the mill or increases in the amount of FFBs processed. The latter can occur as harvesting peaks in the event of a good crop. The decline in aerated lagoon performance is due as much to the increased organic load as well as the increased difficulty experienced in trying to oxygenate the reactor's liquor in the presence of oil.

Both treated POME and PORE can have a very strong brown color (Fig. 10.4.1). This brown coloration is caused by residual organics which are resistant to biological degradation. As noted in Chapter 8, this coloration in treated wastewater is frequently encountered when dealing with agricultural and agro-industrial wastewaters. This coloration has been measured at about 800 Hazens. The brown coloration leads to a residual COD problem in the treated effluent.

Fig. 10.4.1. Although treated PORE has good clarity, it is strongly colored (brown). The BOD and SS of such a treated wastewater can be very low and can easily meet discharge limits of 20:20. The difficulty lies with the residual COD.

PORE treatment, unlike POME treatment, rarely resort to lagooning. This means that the biological unit processes at PORE treatment plants do not have the degree of buffering POME treatment plants have. Variations in wastewater flow and composition have been noted to lead to bioprocess instabilities at PORE treatment plants resulting in treated effluent not being able to meet the discharge limits.

# REFERENCES

## Chapter 1

Ahmed, M.F. & Mohammed, K.N. (1988) Polluting effects of effluent discharges from Dhaka City on the River Buriganga. *Proc. Water Pollution Control in Asia* (eds. T. Panswad, C. Polpraset & K. Yamamoto), pp. 123–129, IAWPRC, Pergamon, UK.

ASEAN/US CRMP (1991) *Technical Publications Series 6: The Coastal Environmental Profile of South Johore, Malaysia.* 65pp., ICLARM, Philippines.

Barril, C.R., Tumlos, E.T. & Moraga, W.C. (1999) Seasonal variations in water quality of Laguna de Bay. *Proc. Asian Waterqual 1999* (eds. C.F. Ouyang, S.L. Lo & S.S. Cheng), pp. 890–895, IAWQ, Taiwan.

Bhuvendralingam, S., DeCosse, P., Liyanamana, P. & Ranawana, S. (1998) Lower Kelani watershed management. *Proc. Water Environment Federation Asia Conference*, Vol. 2, pp. 213–219, WEFTEC, USA.

Chiang, K.M. (1988) River pollution — Clean up and management. *Proc. Water Pollution Control in Asia* (eds. T. Panswad, C. Polpraset & K. Yamamoto), pp. 45–48, IAWPRC, Pergamon, UK.

Du, B. (1995) Coastal and marine environmental management in the People's Republic of China's southern area bordering the South China Sea. *Proc. Coastal and Marine Environmental Management Workshop*, pp. 40–49, Asian Development Bank Publication, Philippines.

Hashimoto, A. & Hirata, T. (1999) Occurrence of *Cryptosporidium* oocysts and *Giardia* cycts in Sagami River, Japan. *Proc. Asian Waterqual 1999* (eds. C.F. Ouyang, S.L. Lo & S.S. Cheng), pp. 956–961, IAWQ, Taiwan.

Jindarojana, J. (1988) Mathematical model: A scientific approach for Nam Pong water quality management. *Proc. Water Pollution Control in Asia* (eds. T. Panswad, C. Polpraset & K. Yamamoto), pp. 29–35, IAWPRC, Pergamon, UK.

Kim, D.I., Cha, D.H., Park, H. & Lee, D.R. (2003) Development of a sustainability assessment strategy for source water conservation in the Han River Basin. *Proc. First International Symposium on Southeast Asian Water Environment*, pp. 467–478, University of Tokyo, Japan.

Liu, C.C.K. & Kuo, J.T. (1988) Wastewater disposal alternatives: Water quality management of Tansui River, Northern Taiwan. *Proc. Water Pollution Control in Asia* (eds. T. Panswad, C. Polpraset & K. Yamamoto), pp. 13–20, IAWPRC, Pergamon, UK.

Matsuo, T. (1999) Japanese experiences in water pollution control and wastewater treatment technologies. *Proc. International Symposium on Development of Innovative*

*Water and Wastewater Treatment Technologies* (eds. G.H. Chen & J.C. Huang), pp. 48–59, HKUST, Hong Kong.

Nguyen, N.S., Hua, C.T., Roop, J., Bansgrove, A., England, S. & McNamee, P.J. (1995) Coastal and marine environmental management in Vietnam. *Proc. Coastal and Marine Environmental Management Workshop*, pp. 50–67, Asian Development Bank Publication, Philippines.

Nguyen, V.A. (2003) Water environment and water pollution control in Vietnam: Overview of status and measure for future. *Proc. First International Symposium on Southeast Asian Water Environment*, pp. 280–287, University of Tokyo, Japan.

Ong, A.S.H., Maheswaran, A. & Ma, A.N. (1987) Chapter 2 — Malaysia. *Environmental Management in Southeast Asia* (ed. L.S. Chia), pp. 14–76, Singapore University Press, Singapore.

Pakiam, J.E., Ch'ng, A.G.S., Mason, W.E., Ng, H.L., Starkey, C., Tam-Lai, S.Y., Tan, G.C. & Tan, K.P. (1980) *Environmental Protection in Singapore*. 114pp, Science Council of Singapore, Singapore.

Sawhney, R. (2003) Environmental challenges of the next decade for Asia and the Pacific region. Presented at *EnviromexAsia2003 — Resource Conservation and Sustainable Development*, Singapore.

*The Straits Times (October 19, 2004) Cancer County.* pp.12, Singapore.

Villavicencio, V.R. (1987) Chapter 3 — Philippines. *Environmental Management in Southeast Asia* (ed. L.S. Chia), pp. 77–108, Singapore University Press, Singapore.

Yassin, A.A.M. (1987) Hazardous wastewater management in the oil and gas industry. *Proc. Conference on Water Management in 2000 for the Developing Countries* (eds. J.H. Tay, K.S. Chong, A. Goh, S.L. Ong, K.S. Periasami & A. Au), pp. WM31–WM43, INTERFAMA, Singapore.

# Index

abdominal pain, 11
acclimated, 64
accumulation, 80
acidic, 54
acidogenic, 73
acidogens, 73
activated carbon, 58, 70
activated sludge, 70, 110, 116, 131, 139, 141
acute, 9
adsorbent, 58
adsorption, 58
aerated lagoon, 93, 116, 131, 139
aeration, 35, 47, 91
aeration pattern, 34
aeration vessel, 32
aerobic, 61
aerobic suspended growth process, 32
aerobically, 34
aesthetic, 10, 56
agglomerate, 62, 80
agro-industrial wastewaters, 2
air blowers, 35
air header, 36
air-water interface, 8
algae, 9
algal blooms, 10
algal growth, 10
alkaline, 54
alkalinity, 49
alum, 49
ambient temperatures, 8
Amm-N, 90
ammonia, 25, 40, 65, 113
ammonium dihydrogen phosphate, 58

ammonium phosphate, 142
amorphous, 49
anaerobic, 61
anaerobic filter, 84
anaerobic lagoon, 74, 116, 131, 138
anaerobic organisms, 9
anaerobic SBR, 117
anaerobic sequencing batch reactor, 81
anaerobically digested, 34
anti-foam agent, 119
anti-vortex, 81, 95
arbitrary flow, 70
*areca catechu*, 124
ASEAN, 3
aspartame, 25
attached growth, 33, 70
attenuation, 92

bacilli, 62
backwashing, 86
bacteria, 61
bacterial solids, 59
baffled tank, 31, 43
bar screens, 43
basket screens, 29
batch reactor, 67
benthic organisms, 8
binary cell division, 63
binding, 29
bioaccumulated, 10
biocatalyst, 59
biochemical oxygen demand (BOD), 65
biodegradability, 116
biodegradable, 5, 114
biodiversity, 9

biofilm, 33, 70, 84, 86
biogas, 76, 78, 81, 86, 117, 138
biological reactor, 62
biomass, 117
biosynthesis, 107
blended, 109
blinding, 29
blood, 125
blowerhouses, 92
BOD:N:P, 11, 58, 137
bowel washing, 125
breweries, 16
buffered, 73
bulking sludge, 64, 109
bund erosion, 142
bunds, 75
butanoic, 73
butanol, 107
butyl acetate, 107

Calcium (Ca), 59
calcium hypochlorite, 36
campaign manufacturing, 21, 106
carbohydrates, 4
carbon dioxide, 73
carbonaceous pollutants, 62
cascade, 68
catalysts, 107
cationic, 83
cell doubling time, 63
cell reproduction, 63
cell residence times or CRTs, 62
cell synthesis, 61
centrifugal forces, 30
centrifuges, 34
chambers, 30, 34, 47
channeling, 84
chemical cracking, 54
chemical oxygen demand (COD), 65
chemical synthesis, 106
chlorination, 111
chlorine, 10
chlorophenols, 10
chronic, 9
circular, 31
clarity, 8, 49

clay, 75
clogging, 85, 91
coagulant, 49
coagulation, 51
coarse air diffusers, 47
coarse screens, 29
coastal waters, 2
Cobalt (Co), 59
coconut, 14
COD:$BOD_5$ ratio, 13
coffee processing, 24
colloidal, 4
color, 52, 142
combined oxygen, 62
complete-mix, 68, 88
composting, 53, 104
concentration profile, 67
contact time, 38
continuous-flow, 67
control panel, 38
conventional activated sludge, 32
conventional complete mix digesters, 77
Copper (Cu), 59
corrosion, 142
countercurrent, 71
counterweights, 79
cover, 78, 81
cow dung, 79
CRT, 89
crude palm oil, 135
crustacea, 62
cryophiles, 64
cryptosporidiosis, 11
*Cryptosporidum*, 3
curved self-cleaning screens, 53
cyclic, 81, 94
cysts, 3

DAFs, 49, 140
dairy product wastewaters, 22
dead zones, 92
decant pump, 81
decant valve, 81
decanters, 95
defeathering, 127
degasifying, 78

dehairing, 125
dehydrate, 64
denatured, 64
denitrification, 61, 91
desalination, 2
desalination plants, 7
desludging, 77, 84, 96
detergents, 24
dewatered, 34
diarrhea, 11
dichromate COD, 66
diffusers, 34, 35, 91, 119
digestion compartment, 80
disinfection, 36
dispersed growth, 62
dissolved air flotation, 130
dissolved air flotator (DAF), 44
dissolved materials, 4
dissolved oxygen, 9
distillery, 15
domestic sewage, 2, 4
dosing pump, 56, 59
downflow, 84
draft tubes, 77
draglines, 77
drum-type mechanical fine screen, 53
drying beds, 34
dyes, 52

earthen construction, 75
*E. Coli*, 11
ecosystems, 2
eddies, 38
eggshells, 30
electricity, 79, 138
electrolytes, 64
energy, 138
environmental degradation, 1
enzyme, 59, 63
equalization, 107, 114
equalization tanks, 46
estuarine, 10
ethanoic acid, 107
ethanol, 107
eutrophication, 10
eviscerating, 125

excavated, 55
excess sludge, 34
explosive, 79
extended aeration, 139

F:M, 88
facultative lagoon, 116
facultative micro-organisms, 62
feces, 113
fermentation, 15, 106, 115
ferric chloride, 49
ferrous sulphate, 49
fertility, 10
fertilizer, 3
filamentous, 63, 109
filter press, 34, 102
fine particulates, 8
fine screen, 15, 53, 129
fish processing, 18
fixed suspended solids (FSS), 8
flexible plastic fibres, 72
flocculation, 51
flow dispersion plate, 84
flow measurement, 29
flow pattern, 4
fluidized bed, 85
foaming, 23, 119
food chain, 10
forced aeration, 71
fouling, 48
fractionation, 134
fresh fruit bunches (FFB), 134
freshwater, 1, 7
fungi, 63

gas injection, 77
gas lances, 77
gas separation, 79
gas yields, 79
gas-solids separator, 80
gastrointestinal disease, 11
*Giardia*, 3
gill surfaces, 8
glaciers, 7
glyceride esters, 134
granular, 80, 82

gravel, 84, 102
gravity flow, 29
gravity settling, 31
gravity thickener, 100
grazers, 109
greasy, 45
grit, 30
grit removal devices, 30
growth phase, 62
growth promoter, 112

Hazens, 142
headworks, 29
heavy metals, 7
helminthes, 11
herbicides, 7
heterotrophs, 61
hexane, 107
high-rate activated sludge, 32
holding tank, 109, 131
homogenous, 67
hoppers, 100
housekeeping, 23
hybrid anaerobic reactor, 86
hydraulic gradient, 29
hydraulic jump, 38
hydraulic loading rates, 32
hydraulic retention times, 32
hydrochloric acid, 55
hydrogen sulphide, 62
hydrolyzed, 73
hydroxypivaldehyde, 107

IDLE, 81, 97
impellers, 29
incinerated, 34
indicator micro-organisms, 11
industrial alcohol, 19
industrial kitchen wastewater, 21
industrial wastewater, 2, 4
industrial wastewater treatment plants, 42
infra-red detector, 66
inhibition, 109
inhibitory, 26, 56
inlet pump sumps, 29
inlet pumps, 29

inline dosing, 38
instantaneous fill, 68
intermittent, 94
ion exchangers, 64
Iron (Fe), 59
iso-propyl alcohol, 107

labyrinthine-type chambers, 38
labyrinthine-type flocculator, 51
landfill, 34, 104
landfill sites, 57
lanolin extraction factory, 47
latex, 57
launder, 31
lift, 29
light penetration, 8
lime, 54, 102
lime powder, 54
liming, 53
lipids, 73
liquid-solids separation, 52
litoral, 10
loading, 79, 80, 84, 102, 116, 117, 138
logarithmic, 55
low speed stirrers, 51
lubricating oil, 47

macerating pumps, 129
macro-nutrients, 58
Magnesium (Mg), 59
magnesium sulphate, 142
malodors, 32
Manganese (Mn), 59
mangrove forests, 2
marco-nutrients, 10
marine environment, 7
marine waters, 10
mechanical mixing, 139
mechanical stirrer, 38
medium-sized, 3
mesocarp, 134
mesophiles, 64
mesophilic, 139
metal activators, 64
metal fragments, 30
metal hydroxide, 49

metal salts, 64
methane, 73
*Methanococcus*, 85
methanogenic, 73
methanol, 107
*methanosarcina*, 80
*methanothrix*, 80
methyl mercury, 2
methyl-isobutyl ketone, 107
micro-nutrients, 59
micro-organisms, 4
microbial population, 62
microbial yield, 63
milk, 19, 49
mineral oil, 8, 46
mixed liquor, 63
mixed liquor volatile suspended solids (MLVSS), 63, 89
MLSS, 63, 97, 100, 139, 141
molasses, 122
molecular oxygen, 62
monosodium glutamate, 24
morphology, 62
moulded plastic shapes, 84
moving bed, 85
moving weir, 95
mud balls, 46
municipal wastewater, 6

nausea, 11
negative pressures, 79, 81
nitrate, 40, 62
nitrification, 61, 91
nitrifiers, 62
nitrogen, 58
nitrogenous, 10
noodle, 20
*Norcadia*, 97
nutrient limiting, 10
nutrient supplementation, 58
nutrients, 58
nutrients removal, 58

O&G traps, 31
obligate aerobes, 62
obligate anaerobes, 73

oil, 134
oil and grease (O&G), 8, 22, 30, 43, 130
olein, 134
oocysts, 3
organic-N, 65
organochlorine pesticides, 7
osmosis, 64
outlet weirs, 31
over-sizing, 51
overflow, 84
oxidation ditch, 88
oxygen dissolution, 8
oxygen demand, 9
oxygen demand gradient, 68
oxygenate, 9

package STP, 39
palm oil, 3, 19
palm oil mill effluent (POME), 75
palm oil refinery effluent (PORE), 136
paper, 16
pathogenic, 4
pathogens, 11
pens, 112, 125
pentanoic, 73
perforated baffle plate, 43
perforated pipes, 47
permanganate COD, 66
persistence, 9
pH adjustment, 54
pharmaceutical, 106
phenol, 10
phosphoric acid, 58
phosphorous, 10, 58
pig, 19
pig farms, 112
piggery wastewater, 3, 112
"pin-head" flocs, 62
pineapple canning wastewater, 53
piperidine acetate, 107
plasmolysis, 64
plastic bags, 29
plug flow, 67
pollution, 1
polyelectrolyte, 102
polymer aids, 51

polymers, 83
POME, 78, 136
porosity, 84
potable water, 2
Potassium (K), 59
potassium dichromate, 65
potassium permanganate, 66
poultry, 13, 19, 125
precipitation, 56
preliminary treatment, 29, 42
pretreatment, 6
primary clarifiers, 32
primary treatment, 31
programmable logic controller (PLC), 84
propanoic, 73
proteins, 4
protozoa, 62
pulp and paper mills, 3

quiescent, 81

racks, 29
rags, 29
rainwater, 114
rat holing, 101
"red-mud" rubber membrane cover, 81
re-oxygenation, 9
reactor, 67
recirculation, 86, 89
red tide, 7, 10
rendering, 132
residual dissolved oxygen, 9
return, 94
rising sludge, 53
rolling motion, 30
rotating, 31
rotating biological contactor (RBC), 71
rotifers, 62
rotor brushes, 88
roughing filter, 71
rubber, 3
rubber gloves factory, 57

saline, 7
sand particles, 30
saponified, 44

sauce making, 24
SBR, 119, 141
scalding, 125
scrapper, 31
screen apertures, 29
scum, 77, 82, 138
seafood processing wastewater, 21
seasonal rainfall, 43
secondary clarifier, 32, 90, 94
secondary treatment, 32
separators, 138
septic, 9, 47, 116
sequencing batch reactor, 131
sequencing batch reactor (SBR) process, 94
settleable suspended solids (SS), 31
settler, 80
settling zone, 80
sewage treatment plants (STPs), 28
sewer network, 29
sheen, 43, 45
shifts, 5
shock load, 76
shock loadings, 86
shortcircuiting, 76
side water depth, 82, 89
silicone, 119
skimmers, 31
slaughterhouse wastewater, 125
slime layer, 62
sludge banks, 8
sludge bed, 79
sludge blanket, 79, 95
sludge cake, 34, 104
sludge digestion, 97, 101
sludge drying beds, 102
sludge layers, 8
sludge treatment, 34
sludge volume index (SVI), 109
slug discharges, 25
slurry, 54
small, 3
soda, 77
sodium chloride, 64
sodium hydroxide, 54
sodium hypochlorite, 36

sodium palmitate, 44
soft drinks bottling plant, 46
soil conditioner, 34, 104
solubility, 8
solvents, 107
spatial frame, 69
spent regenerant, 64
spillages, 134
square clarifiers, 31
stages, 68
standing pig population (spp), 112
start-up, 76, 79, 80, 83
static fine screens, 31
stationary bed, 85
stationary phase, 62
stearin, 134
step-feed, 68
sterilization, 134
stripping, 134
submersible pumps, 43
sugar milling, 19
sulphates, 62
sulphuric acid, 55
support medium, 33, 70, 71
surface aerators, 93
surface area to volume ratio, 33
surface drains, 42
surface overflow rates, 31
surface runoff, 43, 75
suspended growth, 70
sweep-floc, 83
synthetic liners, 75
*Syntrophomonas*, 84

tapered aeration, 68
tapioca, 14
temporal frame, 69
tertiary treatment, 35
textile dyeing, 51
thermal shock, 8
thermophiles, 64

thermophilic, 139
TKN, 113, 128
tobacco, 14
toluene, 107
total dissolved solids (TDS), 64
total organic carbon (TOC), 65
toxic metal sludge, 57
transfer, 91
trapezoidal, 69
trickling filter, 71
TSS (Total Suspended Solids), 5
turbid effluent, 62
two-chambered pH correction
  station, 55

upflow anaerobic sludge blanket (UASB)
  reactor, 79
urea, 58
urine, 4, 113

vacuum, 78
vegetable processing, 21
vermicelli, 20
viruses, 11
viscera, 127
void fraction, 84
volatile fatty acids (VFAs), 73
vomiting, 11
VSS, 80, 82, 84, 101

washout, 84
waste sludge, 34
water seal, 84
weir overflow rate, 31
windbreaks, 76
winery, 19

yields, 99
yoghurt, 19

zinc, 57